KS3 Success

Maths

Practice Test Papers

Age 11-14

Trevor Dixon

KT-432-400

Contents

Introduction

How to Use the Practice Test Papers

About these Practice Test Papers

At the end of Key Stage 3, or Year 9, tests will be used by your teachers to determine your level of achievement in maths.

In this book, there are three sets of test papers that will allow you to track your progress in Key Stage 3 Maths. They will also help you to identify your strengths and weaknesses in the subject.

Sets A, B and C each provide one complete assessment and comprise:
- Test Paper 1: **1 hour** (no calculator allowed)
- Test Paper 2: **1 hour** (calculator allowed)

The test papers will:
- test your knowledge and understanding of the subject, and how you use this knowledge to answer questions
- provide practice questions in all maths topics
- help to familiarise you with the different question styles that appear in test papers
- highlight opportunities for further study and skills practice that will lead to improvement
- record results to track progress.

How to Use the Test Papers

The questions in these test papers have been written in the style that you will see in actual tests.

While you should try to complete the different sections in each set in the same week, you should complete sets A, B and C **at intervals** through Key Stage 3, or Year 9.

Make sure you leave a reasonable amount of time between each assessment – it is unrealistic to expect to see much improvement in just a few weeks. Spreading out the sets will mean you have an opportunity to develop and practise any areas you need to focus on. You will feel much more motivated if you wait for a while, because your progress will be much more obvious.

If you want to re-use the papers, write in pencil and then rub out the answers. However, don't repeat the set too soon or you will remember the answers and the results won't be a true reflection of your abilities.

How to Prepare for the Tests

Revision:
After covering the necessary maths topics, read through your notes from school, or course notes. Perhaps use a revision guide to recap the key points. You could also add notes and diagrams to a mind map.

Equipment you will need:
- pen(s), pencil and rubber
- ruler
- protractor
- pair of compasses
- calculator
- a watch or clock to keep track of the pace at which you are answering questions.

When you feel that you're properly prepared, take the first set of test papers.

Taking the Tests

1. Each set of tests is made up of **two** test papers. Each paper is worth **60 marks**. You should spend **60 minutes** on **each** paper, meaning that one set will take you **two hours**.
2. Check the information about formulae at the end of this introduction.
3. Choose a time to take the first paper when you can work through it in one go. Make sure you have an appropriate place to sit and take the test, where you will be uninterrupted.
4. Answer **all** the questions in the test. If you are stuck on one question, move on and come back to it later. Tests often start with easier questions. The questions become more complex and cover more than one topic as you work through the test papers.
5. Read the questions **carefully**, so that you understand exactly what you need to do. Don't spend too long on any one question.
6. Write the answers in the spaces provided. The space provided for you to write your answer will also give you an indication of how detailed your answer needs to be.
7. The number of marks allocated to each question is shown. This will tell you how many key points the examiner is looking for.
8. Remember that marks may be awarded for key points or working out even if your final answer isn't correct, so **always show your working**, and keep it neat. It may be that if you get the answer wrong, you could still be awarded one mark for showing your working. Sometimes, the second mark for a calculation could be for the **units of measurement**, so make sure you include these.
9. Stay calm! Don't be fazed by questions. Read the question carefully and think it through.

How to Use the Mark Scheme

When you've taken the test, you, or a parent or guardian, should use the mark scheme to mark it. You could mark the test together. It's often helpful for you to discuss the answers with someone as you go through the mark scheme.

The answers and mark scheme will:

- give you an answer to the question **in full**. Any words shown in brackets aren't necessary to obtain the full marks, but should help your understanding of the question
- tell you where alternative answers are acceptable. If it's possible to use different words or terms in an answer, these will be separated by a forward slash, e.g. / . Sometimes when an answer isn't fully correct, certain alternatives may be acceptable
- provide Helpful Hints on answering particular questions.

When you've gone through the test paper, add up the marks to give you your total.

Tips for the Top

After sitting a test paper:

1. Try to analyse your performance. For questions that were incorrect, identify where you went wrong. Are there gaps in your knowledge and understanding? Were there topics where you were under-prepared? Have you misunderstood some of the maths?
2. Pay attention to the Helpful Hints in the answers and mark scheme. These will give you revision tips, and important information about answering a question on a topic. They will also help you to avoid errors made by many students sitting tests.

Check through the following ideas, as they'll help you to do better next time:

1. Make sure your number skills are good – knowing number facts, remembering and using formulae, and being confident with calculation methods will help you to avoid making mistakes.
2. Make sure your knowledge of mathematical facts is accurate – such as knowing metric measures and the meaning of mathematical terms.
3. Check through the paper you have completed.
   - Look for any careless mistakes and try to identify where you went wrong.
   - Look for questions you got wrong – you may not have answered them or you may not have understood.
   - Make a note of the topics and try to focus on these areas to develop your understanding.

Formulae:

You'll be expected to know some formulae, such as those for finding the area and perimeter of rectangles, the area of triangles and parallelograms, and the area and circumference of circles.

Make sure you learn these and can use them. Remember that many questions at Key Stage 3 will ask you to use two of these; for example, finding the area of a compound shape made from a rectangle and a triangle.

We have provided you with two formulae that you'll need to use. Make sure you understand how to use these.

When you've assessed your performance in the first test paper, do any additional work you need to. When you sit the second, and finally the third test, check to see how your performance is improving by comparing marks.

Test Paper 1

Calculator **not** allowed

First name _____

Last name _____

Date _____

Instructions:

- The test is 1 hour long.
- Find a quiet place where you can sit down and complete the test paper undisturbed.
- You **may not** use a calculator for any question in this test.
- You will need: a pen, pencil, rubber and a ruler. You may find tracing paper useful.
- This test starts with easier questions.
- Write your answers where you see this symbol:
- Try to answer all the questions.
- The number of marks available for each question is given in the margin.
- Write all your answers **and working** on the test paper. Marks may be awarded for working.
- Check your work carefully.
- Check how you have done using pages 103–112 of the Answers and Mark Scheme.

You might need to use these formulae:

Trapezium	**Prism**
Area = $\frac{1}{2}(a + b)h$ 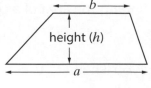 height (h)	Volume = area of cross-section × length 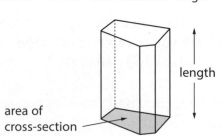 length, area of cross-section

MAXIMUM MARK	60	ACTUAL MARK	

1. Write one of the signs $<$, $>$, $=$ in each circle to complete these statements.

4×-5 ⟨$<$⟩ -4×-5

$5 + -4 + -1$ ◯ $10 - -10$

$-3 - -8$ ◯ $-7 + -4$

$12 \div -4$ ⟨$=$⟩ $-14 \div 7 + -1$

2. Join equivalent numbers with straight lines.

(0.375) $\left(\dfrac{15}{4}\right)$

(3.75) $\left(\dfrac{17}{20}\right)$

(85%) $\left(\dfrac{3}{8}\right)$

3. Calculate the perimeter and area of this shape.

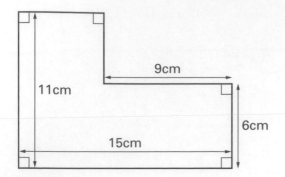

(a) Perimeter:

_____ cm

(b) Area:

_____ cm²

4. Dave and Ali are going to share the cost of a present.

They agree to share the cost in the ratio of 5 : 4.

Dave pays £12.

(a) How much does Ali pay?

£ _____

(b) How much did the present cost?

£ _____

5. A maths group completed a test.

Their scores are:

15	14	14	19	19
20	16	16	13	14

(a) What is the mean score?

_____

1 mark

(b) What is the median score?

_____

1 mark

(c) What is the modal score?

_____

1 mark

(d) What is the range of the scores?

_____

1 mark

SUBTOTAL

1 mark

6. **(a)** Solve:

$$7g + 12 = 21 + g$$

$g =$ _____

1 mark

(b) Simplify this expression.

$$6n + 3p^2 + 5p - p^2 - 4p + 4n$$

_____

1 mark

(c) Find the value of $2a^2 + a^3b$, when $a = 4$ and $b = 5$.

_____

2 marks

7. Find the volume of this cylinder.

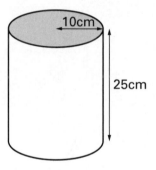

10cm

25cm

Use π = 3

Give your answer as a whole number.

_____ cm^3

8. **(a)** Express 24 as a product of its prime factors.

1 mark

(b) Find the highest common factor of 36 and 96.

1 mark

(c) Find the lowest common multiple of 15 and 24.

1 mark

9. A regular polygon has 12 sides.

(a) Lines are drawn from one vertex to other vertices to make triangles. **Some** lines have been drawn for you.

How many triangles will there be altogether?

1 mark

(b) Find the sum of its interior angles.

1 mark

_____°

(c) Find the size of each of its exterior angles.

1 mark

_____°

SUBTOTAL

10. There are some red, blue, black and white counters in a bag.

There are 12 blue counters.

One counter is selected at random. The probability of it being a certain colour is as follows:

- A red counter 0.2

- A blue counter 0.4

- A black counter 0.1

(a) Explain why it is impossible for there to be 25 counters in the bag.

_____

_____

_____

(b) What is the probability of taking a white counter from the bag?

_____

(c) How many white counters are in the bag?

_____

11.   *ABC* and *DEF* are similar triangles.

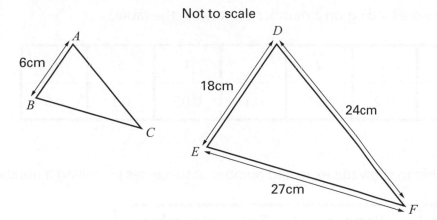

(a) What is length of the side *BC*?

_____ cm

1 mark

(b) A third similar triangle is drawn by increasing *DEF* by a scale factor of 4.

Write the lengths of the sides of the third similar triangle.

_____ cm, _____ cm, _____ cm

2 marks

SUBTOTAL

12. Sally trials rolling a die.

The probability of the die landing on a number is shown in the table.

Number on die	1	2	3	4	5	6
Probability	0.2	0.05	0.1	0.05	0.1	0.5

2 marks

(a) Complete this table to show the expected number of successes for rolling a number.

Number rolled on die	Number of trials	Expected number of successes
1	25	_____
2	40	_____
3	90	_____

1 mark

(b) Sally says, 'The dice is fair because any number can be rolled.'

Is Sally correct?

Circle YES or NO.

YES / NO

Explain your answer.

_____

_____

_____

13. This frequency table shows the scores in a maths test.

Score	Frequency	Score × frequency
20	1	_____
19	2	_____
18	3	_____
17	7	_____
16	4	_____
15	3	_____
Totals	_____	_____

(a) Complete the table.

3 marks

(b) Find the median number on this stem and leaf diagram.

3	2	6	7		
4	6	8			
5	1	5	8	9	
6	2	2	5	7	9
7	8				

1 mark

_____

SUBTOTAL

14. **(a)** Draw lines to join the inverse functions.

2 marks

$x \rightarrow 2x + 3$ $y \rightarrow \dfrac{y - 2}{3}$

$x \rightarrow 2x - 3$ $y \rightarrow \dfrac{y + 2}{3}$

$x \rightarrow 3x + 2$ $y \rightarrow \dfrac{y - 3}{2}$

$x \rightarrow 3x - 2$ $y \rightarrow \dfrac{y + 3}{2}$

1 mark

(b) Give the inverse function of this equation in the form $y =$

$$\dfrac{(x + 5)}{4} \qquad \rightarrow$$

 $y = $ _____

15. Sally and Yasmin like to make a drink from a mixture of pear and apple juice.

Sally likes to mix the pear and apple juice in the ratio of 2 : 3.

Yasmin prefers the mixture to have the ratio of 3 : 2.

1 mark

(a) What fraction of Sally's drink is made from apple juice?

 _____

1 mark

(b) Yasmin uses 450 ml of apple juice.

What is the difference between the amount of apple and pear juice in Yasmin's drink?

Give your answer in millilitres (ml).

 _____

2 marks

(c) Sally makes 1 litre of her mixture.

She drinks half of her juice mixture.

How many millilitres of pear juice will Yasmin have to add to the rest so it makes her mixture?

 _____

16. The lengths of four sides on this hexagon are identified.

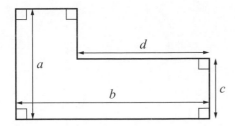

(a) Tick (✓) the expression that does **not** represent a way of finding the area of the hexagon.

1 mark

(i) ☐ $ab - d(a - c)$

(ii) ☐ $a(b - d) + cd$

(iii) ☐ $a(b - d) + bc$

(iv) ☐ $(a - c) \times (b - d) + bc$

(b) Write a formula to find the length b when the perimeter P and width a are known.

1 mark

$b = \underline{\hspace{3cm}}$

SUBTOTAL

17. AB is the diameter of the circle and is 20cm.

Use π = 3

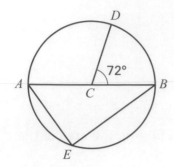

(a) Calculate the length of the arc BD.

 _____ cm

(b) Calculate the area of the sector BCD.

Give your answer as a whole number.

 _____ cm²

(c) $\angle ABE = 38°$

Calculate $\angle BAE$.

 _____ °

18. Cuboids A, B and C are similar.

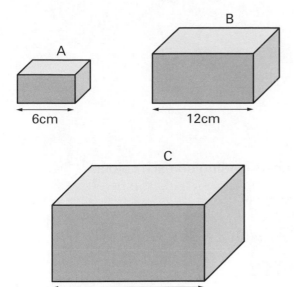

The ratio of length : width : height is 3 : 2 : 1 in these cuboids.

(a) Calculate the ratio volume of A : volume of B.

_____ : _____

2 marks

(b) Calculate the ratio volume of A : volume of C.

_____ : _____

2 marks

SUBTOTAL

19. These spinners are identical.

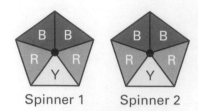

Spinner 1   Spinner 2

The chart shows the probability of the spinners landing on each colour.

Yellow	Red	Blue
0.1	0.5	0.4

Calculate these probabilities.

(a) The probability of spinning yellow on both spinners.

(b) The probability of not spinning yellow on both spinners.

(c) The probability of spinning red on Spinner 1 and blue on Spinner 2.

1 mark

1 mark

1 mark

20. Convert these numbers to standard form.

(a) $(3 \times 10^5) \times (5 \times 10^3)$

1 mark

(b) $(6 \times 10^7) \div (4 \times 10^3)$

1 mark

21.

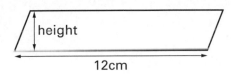

height

12cm

The area of this parallelogram is 30 cm².

The length is 12 cm.

What is the height of the parallelogram?

1 mark

SUBTOTAL

Test Paper 2

Calculator allowed

First name _____

Last name _____

Date _____

Instructions:

- The test is 1 hour long.
- Find a quiet place where you can sit down and complete the test paper undisturbed.
- You **may** use a calculator for any question in this test.
- You will need a pen, pencil, rubber, ruler, a pair of compasses and a scientific or graphic calculator.
- This test starts with easier questions.
- Write your answers where you see this symbol:
- Try to answer all the questions.
- The number of marks available for each question is given in the margin.
- Write all your answers **and working** on the test paper. Marks may be awarded for working.
- Check your work carefully.
- Check how you have done using pages 103–112 of the Answers and Mark Scheme.

You might need to use these formulae:

Trapezium	**Prism**
Area = $\frac{1}{2}(a + b)h$	Volume = area of cross-section × length

MAXIMUM MARK	60		ACTUAL MARK	

1. **(a)** Write the next three terms in this sequence.

19   12   5   _____   _____   _____

1 mark

(b) To find the next number in this sequence, double the number and subtract 4.

Write the next three terms.

5   6   8   _____   _____   _____

1 mark

(c) To find the next term in this sequence, follow this rule: $3n - 7$

Write the next three terms.

−4   −1   2   _____   _____   _____

1 mark

2. A fruit dessert is made using 200 g of oranges, 120 g of apples and 160 g of pears.

(a) What is the ratio of oranges to apples to pears?

Write the ratio in its simplest terms.

_____ : _____ : _____

1 mark

(b) What fraction of the dessert is apple?

Write the fraction in its simplest terms.

_____

1 mark

SUBTOTAL

3. This bag contains coloured counters.

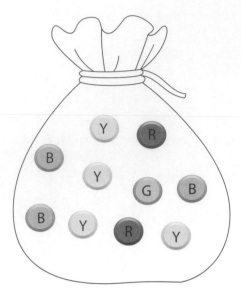

(a) What is the probability that a counter taken from the bag will **not** be yellow?

(b) A yellow counter and a green counter are taken from the bag.

What is the probability that the third counter taken from the bag will be blue?

(c) Five black counters are added to the original bag of counters.

What is the probability that a blue counter will be taken from the bag?

1 mark

1 mark

1 mark

4. Here is a train timetable.

London King's Cross	14:00	14:35	15:05	15:40	16:05	16:25	16:55
Peterborough		15:36	15:58			17:18	17:49
Doncaster	15:35	16:22				17:39	
York	16:27	16:57	17:23	17:35		18:35	
Newcastle	17:31	18:04	18:19		19:02	19:32	19:48
Edinburgh		19:42	19:58	20:12		20:55	21:20

(a) Ewan arrives at London King's Cross station at 4pm to catch a train to Newcastle.

When should he arrive in Newcastle?

 _____

1 mark

(b) Milly has to be in York for six o'clock in the evening.

Which would be the best train to catch from Peterborough?

 _____

1 mark

(c) Dev catches the 14:00 from London King's Cross. He gets off in York and spends one hour meeting someone at York Station. He then gets the next train to Newcastle.

When will he arrive in Newcastle?

 _____

1 mark

SUBTOTAL

5. A cuboid measures 12cm × 8cm × 4cm.

(a) How many edges will be 12cm long?

1 mark

_____

(b) What will the surface area of the cuboid be?

2 marks

_____ cm^2

6. The curved lines in this shape are the circumferences of two semi-circles joined by two straight lines.

Give your answers correct to 2 decimal places.

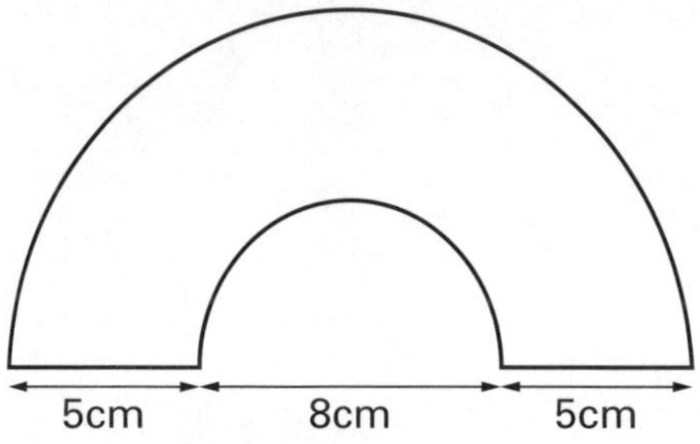

| 5cm | 8cm | 5cm |

(a) Find the area of the shape.

2 marks

_____ cm^2

(b) Find the perimeter of the shape.

2 marks

_____ cm

7. Multiply out the brackets in these expressions.

(a) $4(s^2 - 5s)$

1 mark

(b) $(r + 6)(r + 7)$

1 mark

Factorise.

(c) $10t + 25$

1 mark

8. **(a)** 60% of a number is 360.

What is the number?

1 mark

(b) Jack spent $\frac{3}{8}$ of his savings on buying DVDs.

He spent £19.50

How much of his savings did he have left?

1 mark

SUBTOTAL

9. This table shows the numbers of jumpers on sale in a shop.

The sizes and colours of the jumpers are shown.

		Colour			
		Brown	Blue	Green	Red
Size	Extra large	2	4	4	1
	Large	3	3	6	5
	Medium	4	6	5	5
	Small	3	4	3	2

(a) A jumper is chosen at random.

What is the probability that the jumper chosen will be brown?

(b) The brown jumper is replaced. A jumper is chosen at random.

What is the probability that the jumper chosen will be medium?

(c) Martha thinks about picking a jumper at random.

She says, 'Picking a large, red jumper is a mutually exclusive event.'

Is Martha correct? Circle YES or NO   YES / NO

Explain your answer.

 _____

_____

_____

10. This table shows the cost of sending letters and parcels by post.

Letters				
	First class		Second class	
	Letter	Large letter	Letter	Large letter
0–100 g	60p	90p	50p	69p
101 g–250 g		£1.20		£1.10
251 g–500 g		£1.60		£1.40
501 g–750 g		£2.30		£1.90
Parcels				
	First class		Second class	
	Small parcel	Medium parcel	Small parcel	Medium parcel
Up to 1 kg	£3.00	£5.65	£2.60	
1 kg–2 kg	£6.85	£11.90	£5.60	£8.90
2 kg–5 kg		£15.10		£12.92
5 kg–10 kg		£21.25		£15.92
10 kg–15 kg		£32.40		£22.46
15 kg–20 kg		£32.40		£27.68
20 kg–25 kg				£38.48
25 kg–30 kg				£42.50

SUBTOTAL

1 mark

(a) Mr Jones sends the following:

- 2 first-class letters.

- 3 second-class large letters weighing 300g each.

- 1 first-class medium parcel weighing 8kg.

What does Mr Jones have to pay?

 £ _____

2 marks

(b) Mrs Singh pays £26.40 for the following:

- 4 second-class large letters weighing 400g.

- 2 medium parcels.

Circle the prices she paid for the 2 medium parcels.

	First class	**Second class**
	Medium parcel	Medium parcel
Up to 1kg	£5.65	
1kg–2kg	£11.90	£8.90
2kg–5kg	£15.10	£12.92
5kg–10kg	£21.25	£15.92
10kg–15kg	£32.40	£22.46
15kg–20kg	£32.40	£27.68
20kg–25kg		£38.48
25kg–30kg		£42.50

11. Solve:

(a) $\dfrac{5x + 13}{6} = 8$

2 marks

$x = \underline{\hspace{2cm}}$

(b) $\dfrac{4x - 2}{3} = \dfrac{3x + 9}{4}$

2 marks

$x = \underline{\hspace{2cm}}$

12. Use a calculator to solve the following.

Round your answer to two decimal places.

1 mark

$$\dfrac{35.78\,(56.7 - 7.08)}{3.4 + 2.3 \times 8.32}$$

$\underline{\hspace{2cm}}$

13. Look at this sequence.

1 mark

5   8   13   20   29

Tick (✓) the expression that describes the nth term in the sequence.

$\square$ $n^2 + 3$

$\square$ $n^2 + 4$

$\square$ $2n + 3$

$\square$ $3n + n^2$

$\square$ $2n^2 + 3$

SUBTOTAL

1 mark

14. Find two pairs of quadrilaterals that will join to make two similar rectangles.

Not to scale

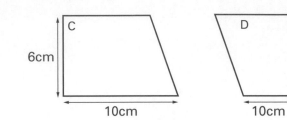

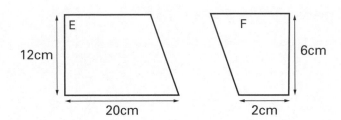

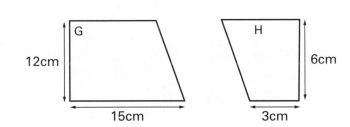

_____ and _____ make a rectangle similar to _____

and _____

15. Tom wants to buy a game console.

The full price is £225

The price is reduced to £180

(a) What is the percentage decrease?

1 mark

_____

Tom buys the game console at the reduced price. He pays a deposit of £30

The shop adds a percentage to the balance left to pay.

He decides to pay the rest in 12 instalments.

Tom has to pay 12 instalments of £13.25

(b) What is the percentage increase the shop has added to pay by instalments?

2 marks

_____

The shop also has computer games for sale.

Each game costs £40

The price for a pack of three games is £72

(c) What is the percentage saving when buying a pack of three games instead of buying three separate games?

2 marks

_____

SUBTOTAL

16.

1 mark

(a) Here are 7 number cards. They are all integers.

| 12 | 15 | 8 | 20 | 34 | ? | ? |

The mean of the seven numbers is 16.

What could the other two numbers be?

_____ and _____

1 mark

(b) Here are 7 number cards. They are all integers.

| 13 | 15 | 6 | 12 | 7 | 20 | ? |

The range of the numbers is 18.

What could the other number be?

_____ and _____

1 mark

(c) Here are 7 number cards. They are all integers.

| 12 | 10 | 9 | 9 | 10 | ? | ? |

The mode of the numbers is 9, the median is **not** 9 and the range is 3.

What are the other two numbers?

_____ and _____

17. **(a)** Here are two expressions.

$4a + 3$ $\qquad\qquad$ $5a - 1$

What value of a would make the expressions equal?

 $a =$ _____

2 marks

(b) Solve these simultaneous equations.

$5b - 2c = 17$

$8b + 2c = 48$

 $b =$ _____ and $c =$ _____

2 marks

(c) The lengths of the triangle are shown.

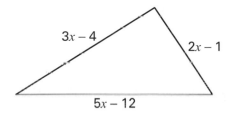

(i) Write an expression for the perimeter (P) of the triangle.

 $P =$ _____

1 mark

(ii) The perimeter of the triangle is 53cm.

Find the lengths of the three sides.

 The sides are _____ cm, _____ cm and _____ cm

1 mark

SUBTOTAL

18. The options for a sports lesson are shown in this two-way table.

Sports	Male	Female	Total
Football	17	15	32
Basketball	15	13	28
Swimming	14	16	30
Total	46	44	90

(a) A boy is chosen at random. What is the probability that he is playing basketball?

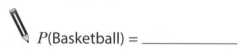 P(Basketball) = _____

(b) A pupil is chosen at random. What is the probability that they are swimming?

P(Swimming) = _____

(c) When it rains, football is played in the Sports Hall.

P(raining) = 0.2

Complete the tree diagram to calculate the probability it will **not** rain on two consecutive days.

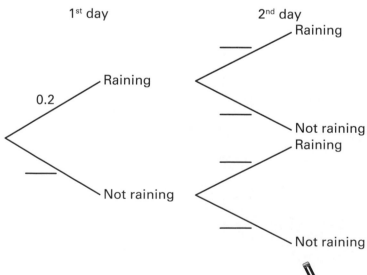

P(no rain for 2 days) = _____

36

19. A school carried out a survey to find how far children lived from school.

There are 600 students in the school.

The data is presented in a cumulative frequency graph.

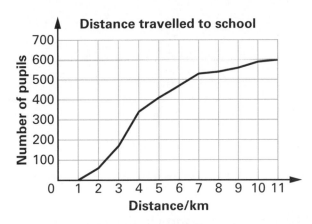

Use the graph to work out:

(a) the median distance travelled to school.

1 mark

(b) the interquartile range for the school's data.

2 marks

20. Simplify the following:

(a) $6d^{-3} \times 3d^{-2}$

1 mark

(b) $24c^{7} \div 8c^{-2}$

1 mark

(c) $\dfrac{10a^3b^4}{2ab^2}$

1 mark

SUBTOTAL

Test Paper 1

Calculator **not** allowed

First name _____

Last name _____

Date _____

Instructions:

- The test is 1 hour long.
- Find a quiet place where you can sit down and complete the test paper undisturbed.
- You **may not** use a calculator for any question in this test.
- You will need: a pen, pencil, rubber and a ruler. You may find tracing paper useful.
- This test starts with easier questions.
- Write your answers where you see this symbol:
- Try to answer all the questions.
- The number of marks available for each question is given in the margin.
- Write all your answers **and working** on the test paper. Marks may be awarded for working.
- Check your work carefully.
- Check how you have done using pages 103–112 of the Answers and Mark Scheme.

You might need to use these formulae:

Trapezium	Prism
Area = $\frac{1}{2}(a + b)h$	Volume = area of cross-section × length

MAXIMUM MARK	60		ACTUAL MARK	

1. Tim needs to cut a 4m plank of wood into five equal lengths to make some shelves.

(a) How long will each shelf be?

1 mark

_____

(b) Tim is going to use the shelves for his DVDs.

Each DVD is 1.5cm wide.

How many DVDs will fit on each shelf?

2 marks

_____

2. Calculate 683 ÷ 17.

2 marks

Give your answer correct to two decimal places.

_____

SUBTOTAL

3. These two triangles are congruent.

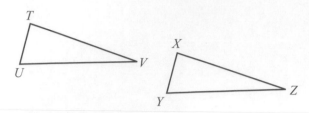

$\angle TUV = 75°$

$\angle YXZ = 87°$

(a) Calculate $\angle XZY$

 _____ °

(b) Explain why TUV and XYZ are **not** isosceles triangles.

_____

_____

_____

(c) There are two possible isosceles triangles that could have at least one angle of 76°.

Write the three angles of each of these triangles.

 _____ °, _____ ° and _____ °

and

 _____ °, _____ ° and _____ °

40

4. **(a)** The lengths of the sides of this rectangle are shown by expressions.

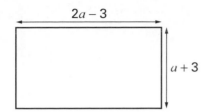

$2a - 3$

$a + 3$

Write an expression for the area (A) of the rectangle.

 $A =$ _____

1 mark

(b) This design is made from nine rhombuses.

The length of each side of a single rhombus is $2b + 1$.

Find the perimeter (P) of the design if $b = 8$ cm.

 $P =$ _____

1 mark

SUBTOTAL

5. This pie chart shows how 720 children travel to school.

Travelling to school

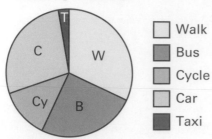

☐ Walk
◼ Bus
◼ Cycle
☐ Car
◼ Taxi

(a) Estimate the number of children who get the bus to school.

(b) Which means of transport represents about 12.5% of the children?

6. This bar chart shows the weekly wages of the staff at an office.

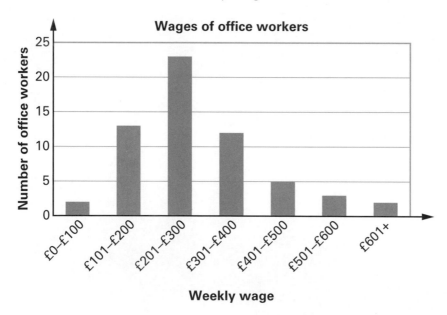

(a) How many office workers earn between £101 and £300 a week?

The office manager wants to make a pie chart of this information.

(b) What fraction of the pie chart would be shown by the workers who earn between £401 and £500 a week?

1 mark

_____

(c) What would be the angle at the centre of the sector for the workers who earn between £301 and £400 a week?

1 mark

_____ °

7. A map has a scale of 1 : 50 000.

(a) A walk is 4 kilometres long. How long is the walk on the map?

Give your answer in centimetres (cm).

2 marks

_____ cm

(b) On the map, two towns are 25 cm apart. How far are the towns apart in real life?

Give your answer in kilometres (km).

2 marks

_____ km

8. A number is divided by 26 and the answer is 43.6

What is the number?

1 mark

_____

SUBTOTAL

2 marks

9. Nisha has two fair 1 to 6 dice.

She asks herself these questions about the trials she is trying.

Tick (✓) the correct statement for each question.

(a) 'How can I describe the events of "rolling a 6" with each dice?'

These are independent events.	These are mutually exclusive events.	These are dependent events.

(b) 'What is the probability of totalling both scores to reach 12?'

The probability is $\frac{1}{6}$	The probability is $\frac{1}{12}$	The probability is $\frac{1}{36}$

(c) 'What is the probability of rolling one die and the number being < 2 and even?'

This is an independent event.	This is a mutually exclusive event.	This is a dependent event.

(d) 'What is the probability of rolling a number < 3 on one dice and an even number on the second dice?'

The probability is $\frac{1}{2}$	The probability is $\frac{1}{5}$	The probability is $\frac{1}{6}$

10. Mark, Barry and Carla have some marbles.

Barry has three times as many marbles as Mark.

Carla has four times as many marbles as Mark.

(a) Write a formula to show the total number of marbles, M, and let the number of Mark's marbles be x.

$M =$ _____

1 mark

(b) Carla has 140 marbles.

How many marbles do Mark and Barry have?

Mark has _____ marbles and Barry has _____ marbles.

1 mark

11. Manisha completed a test. She got 17 questions out of 20 right. Her teacher wrote her score as a percentage.

(a) What was her percentage score?

_____

1 mark

(b) Manisha's friend Sally scored 70%.

How many out of 20 did Sally score?

_____

1 mark

(c) Dev completed a different test. He got 18 out of 25 questions right.

Dev says, 'I scored more than Manisha, so I did best.'

Manisha says, 'No, I did best.'

Explain why Manisha is correct.

_____

_____

_____

1 mark

SUBTOTAL

12. Garden fertiliser uses a ratio to show the amount of nitrogen (N), phosphorus (P) and potassium (K) it contains. It is called the NPK ratio.

A 6 : 4 : 4 ratio means the fertiliser contains 6% nitrogen, 4% phosphorus and 4% potassium. The remaining percentage is a filler.

(a) A pack of fertiliser has an NPK ratio of 25 : 2 : 8.

The pack has 160g of potassium.

What is the mass of the full pack, including the filler?

Give your answer in kilograms (kg).

_____ kg

(b) There is 200g **each** of nitrogen, phosphorus and potassium.

How much fertiliser can be made if the fertiliser has an NPK ratio of 25 : 2 : 8?

Give your answer in kilograms (kg).

_____ kg

13. By rounding to 1 significant figure, estimate the answers to these calculations.

(a) 831×0.709

_____

(b) $(4811 + 4427) \div 27$

_____

14. There are three pairs of triangles.

Tick (✓) the pairs that are congruent and give the condition of congruency you are using.

(i)

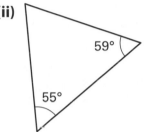   

☐   🖉 _____

(ii)

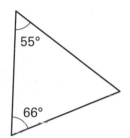

☐   🖉 _____

(iii)

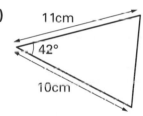   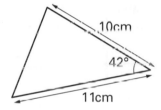

☐   🖉 _____

15. A nursery teacher has a set of shapes.

All the shapes are pentagons. There are three different sizes:

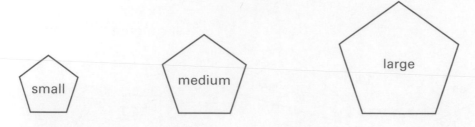

small   medium   large

Each of the sizes comes in four different colours:

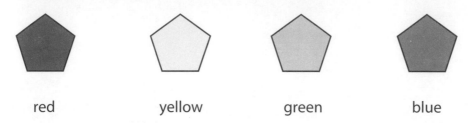

red   yellow   green   blue

The teacher takes **some** of the shapes.

The probability of picking the shapes the teacher selected at random is listed in the table.

Pentagon	Probability	Pentagon	Probability
Small red	0.1	Small green	0.05
Medium red	0.125	Medium green	0.15
Small yellow	0.25	Large green	0.125
Large yellow	0.15	Small blue	0.05

(a) The probability of selecting a medium red pentagon is 0.125

There are 10 medium red pentagons.

How many pentagons did the teacher select altogether?

 _____

(b) The teacher did not select every type of pentagon.

Use the table to identify the types of pentagon the teacher did **not** select.

16. **(a)** Solve:

$$3(5x + 3) + 4(3x − 3) = 105$$

 $x =$ _____

(b) Solve this pair of simultaneous equations.

$$5x + 2y = 26$$

$$2x + 2y = 14$$

$x =$ _____ , $y =$ _____

(c) Make f the subject of this equation.

$$e + 4 = \frac{d^2}{f} + 7$$

$f =$ _____

17. The mean of five numbers was 21.

Here are some facts about the five numbers:

- The largest number was 40.
- The median number and the mode were both 16.
- The range was 28.

(a) When the five numbers were arranged in order, from lowest to highest, what was the fourth number?

_____

A sixth number was added to the group.

The new mean was 24.

(b) What was the sixth number added to the group?

 _____

18. *ABG*, *ACF* and *ADE* are all similar triangles.

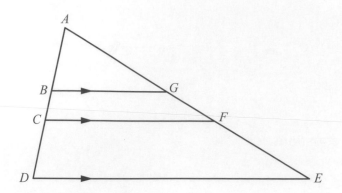

- Line *AG* is 10cm

- Line *AD* is 15cm

- Line *BG* is 8cm

- Line *DE* is 20cm

- Line *EF* is 10cm

Calculate the lengths of:

(a) *GF*

_____ cm

(b) *CF*

_____ cm

(c) *BC*

_____ cm

19. This is an octagonal prism.

It is 15cm long.

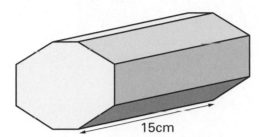

The end face is not a regular octagon, but does have two lines of symmetry.

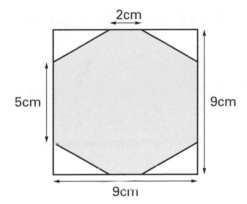

2cm

5cm   9cm

9cm

(a) Calculate the area of the end face of the octagonal prism.

_____ cm²

(b) Calculate the volume of the prism.

_____ cm³

20. 125 students were asked about how they spent the previous evening.

Their answers were recorded in a sorting diagram.

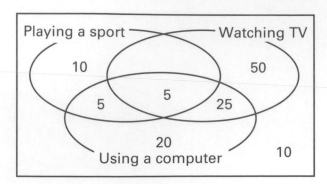

(a) What is the probability that a student chosen at random will have been using a computer?

(b) What is the probability that a student chosen at random will not have watched TV?

The same group of students was surveyed again one week later.

This time the probability that a student did not take part in any activity was 0.04

(c) How many students did not take part in any activity?

(d) Explain why the probability of a student in the second survey using a computer could not be 0.1

52

21. The scatter graph shows the relationship between the height and shoe size of some students.

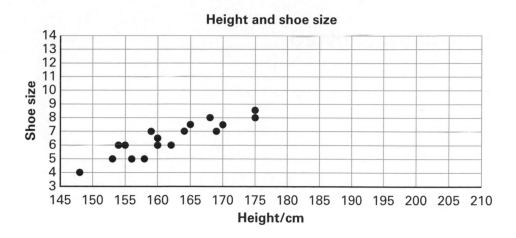

(a) Describe the correlation of the graph.

_____

1 mark

(b) On the graph, draw a line of best fit.

1 mark

(c) Estimate the shoe size for someone who is 195cm tall.

_____

1 mark

(d) Holly says, 'This graph shows that all students who are 160cm tall would never wear shoes that are size 10.'

1 mark

Is Holly correct?

Circle YES or NO   YES / NO

Explain your answer:

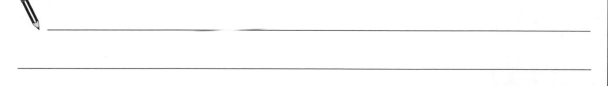

_____

_____

_____

Test Paper 2

Calculator allowed

First name _____

Last name _____

Date _____

Instructions:

- The test is 1 hour long.
- Find a quiet place where you can sit down and complete the test paper undisturbed.
- You **may** use a calculator for any question in this test.
- You will need a pen, pencil, rubber, ruler, a pair of compasses and a scientific or graphic calculator.
- This test starts with easier questions.
- Write your answers where you see this symbol:
- Try to answer all the questions.
- The number of marks available for each question is given in the margin.
- Write all your answers **and working** on the test paper. Marks may be awarded for working.
- Check your work carefully.
- Check how you have done using pages 103–112 of the Answers and Mark Scheme.

You might need to use these formulae:

Trapezium	**Prism**
Area $= \frac{1}{2}(a + b)h$	Volume = area of cross-section × length
b height (*h*) *a*	length, area of cross-section

MAXIMUM MARK	60		ACTUAL MARK	

1. Calculate the missing angles.

(a) Find ∠r

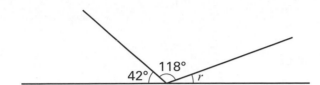

1 mark

r = _____ °

(b) Find ∠s

1 mark

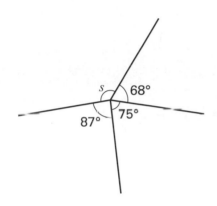

s = _____ °

(c) Find ∠t

2 marks

t = _____ °

SUBTOTAL

2. Glen's music collection is made up from CDs and downloads.

He has 5 downloads for every 7 CDs.

Glen has 35 CDs.

1 mark

(a) How many downloads does he have?

_____

Glen's friend Ben has twice as many CDs, but the same number of downloads.

1 mark

(b) What is the ratio of Ben's CDs to downloads in its simplest terms?

_____ : _____

3. A bag has 5 counters; 3 of the counters are black and 2 are white.

1 mark

(a) A counter is drawn at random; what is the probability that the counter is black?

Give your answer in its simplest terms.

_____

1 mark

(b) A black counter is drawn and is not replaced in the bag.

What is the probability that the next counter drawn at random will be black?

_____

4. Here are six number cards.

The sixth card is turned over.

| 4 | 4 | 4 | 4 | 4 | |

The mean of the six numbers is 5.

(a) What is the number on the sixth card?

_____

Here are another six number cards, again the sixth card is turned over.

| 6 | 3 | 7 | 4 | 7 | |

The range of the six numbers is 6.

(b) What are the two numbers that the sixth card could be?

_____ or _____

5. **(a)** Find the value of $\dfrac{6a + 12}{a}$ when $a = 1.5$

_____

(b) Solve:

$2b + 6 = 5c + 8$ when $c = 0.5$

$b =$ _____

1 mark

1 mark

1 mark

1 mark

6.   **(a)** Find $\frac{4}{5}$ of £28.

_____

(b) £28 is $\frac{4}{5}$ of an amount.

What is the amount?

_____

7.   Using a ruler and compasses:

(a) construct the perpendicular bisector of this line.

A ——————————————— B

9cm

(b) construct the bisector of this angle.

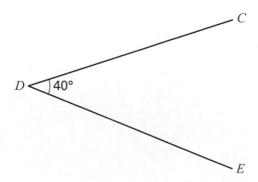

C

D ◁ 40°

E

8. Barry has £75 in his bank account.

He takes £30 from the account.

(a) Calculate the ratio, in its simplest terms, of the money he has taken out to the money he has left in his account.

_____ : _____

2 marks

Barry decides to spend $\frac{1}{3}$ of the money.

(b) Re-calculate the ratio, in its simplest terms, of the money he now has left to the money left in his account.

_____ : _____

2 marks

9. Calculate:

(a) $5\frac{2}{3} + 6\frac{7}{8} =$

_____

1 mark

(b) $6\frac{2}{5} - 3\frac{2}{3} =$

_____

1 mark

SUBTOTAL

10.

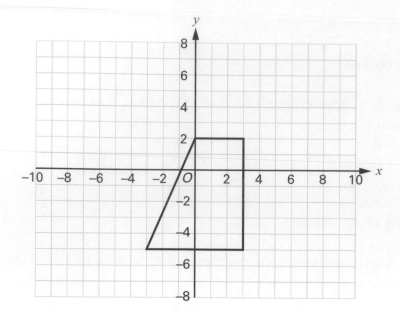

(a) What are the coordinates of the quadrilateral?

     (_____ , _____), (_____ , _____),

(_____ , _____), (_____ , _____)

(b) The quadrilateral is reflected in the y-axis.

What are the new coordinates?

     (_____ , _____), (_____ , _____),

(_____ , _____), (_____ , _____)

11. This is a triangular spinner.

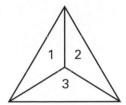

It is a biased spinner.

Spinning a 2 is twice as likely as spinning a 1.

Spinning a 3 is twice as likely as spinning a 2.

What is the probability of spinning a 1?

2 marks

12. The ages of members of a gym club are shown.

	Ages														
Male	21	32	26	21	34	45	46	19	53	24	33	40	61	49	53
Female	24	27	39	40	18	27	45	61	26	35	37	49	34	21	62

Complete this two-way table to show the frequencies.

	Male	Female
$10 < A \leqslant 20$		
$20 < A \leqslant 30$		
$30 < A \leqslant 40$		
$40 < A \leqslant 50$		
$50 < A \leqslant 60$		
$60 < A \leqslant 70$		

2 marks

SUBTOTAL

13. Four circles fit exactly into this square.

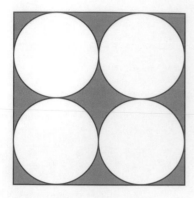

The area of the square is 625cm².

(a) Find the area of one circle.

Give your answer correct to 2 decimal places.

 cm²

(b) Calculate the shaded area.

 cm²

14. Factorise:

(a) $5x - 25$

(b) $4ab^2 + 2bc$

Area = $12d^2$ $4d$

(c) Write a simplified expression for the length of this rectangle.

15. A cake has this nutritional information.

(a) Complete the missing values in the table.

Write the values to 1 decimal place.

	Per 120g	Per 50g
Protein	9g	3.75g
Carbohydrates	79.4g	_____ g
of which sugars	_____ g	8.7g
Fat	2.1g	_____ g
of which saturates	0.6g	0.25g
Fibre	_____ g	1.8g

(b) The cake weighs 900g.

(i) How much protein does it contain?

_____ g

(ii) How much saturated fat does it contain?

_____ g

16. Milly has a money box with 60 coins.

There are no 2p or 1p coins.

This table shows the probabilities of picking a coin at random.

Coin	Probability
£1	0.3
50p	0.25
20p	0.2
10p	0.1
5p	0.15

(a) Explain why it is impossible that Milly has any £2 coins in her money box.

_____

_____

_____

1 mark

(b) Calculate how much money Milly has altogether.

£ _____

2 marks

17. This table shows the results of a school maths test.

Score	Frequency
$\leqslant 20\%$	2
$> 20\% \leqslant 40\%$	4
$> 40\% \leqslant 60\%$	8
$> 60\% \leqslant 80\%$	12
$> 80\% \leqslant 100\%$	6

(a) What fraction of the scores were in the $> 20\% \leqslant 40\%$ category?

Give your answer as a simplified fraction.

1 mark

(b) Use this circle to display the information from the table in a pie chart.

2 marks

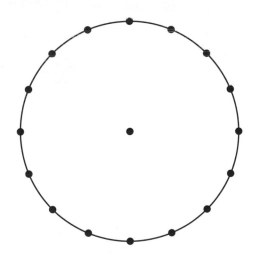

(c) Calculate the angle at the centre of the sector for the $> 60\% \leqslant 80\%$ category.

 °

1 mark

SUBTOTAL

1 mark

18. **(a)** Here are possible lengths for a triangle.

　　　5cm　　　　6cm　　　　7cm

　　　8cm　　　　9cm　　　　10cm

Which three lengths could be used for a right-angled triangle?

 _____ cm, _____ cm and _____ cm

2 marks

(b)

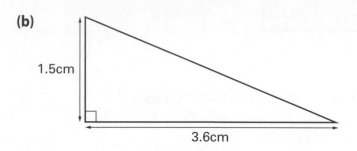

1.5cm

3.6cm

Calculate the area of this triangle.

 _____ cm²

1 mark

(c) The square of the hypotenuse of a right-angled **isosceles** triangle is 800cm².

Calculate the length of one of the other sides.

_____ cm

19. **(a)** Write the first 3 numbers in the sequence $5n^2 - 4$.

1 mark

_____, _____, _____

(b) Write the expression that is shown by the curve on the graph.

1 mark

_____

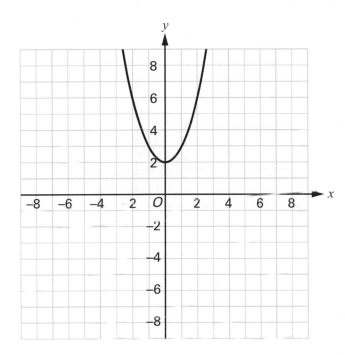

(c) On the graph, draw a curve representing the equation $y = x^2 - 3$.

1 mark

SUBTOTAL

20. The amount of money raised during a charity fun run is recorded in a table.

Amount raised per person (n)	Frequency	Cumulative frequency
$0 < n \leqslant 50$	5	
$50 < n \leqslant 100$	12	
$100 < n \leqslant 150$	28	
$150 < n \leqslant 200$	20	
$200 < n \leqslant 250$	18	
$250 < n \leqslant 300$	12	
$300 < n \leqslant 350$	4	
$350 < n \leqslant 400$	1	

(a) Draw a cumulative frequency graph of this data.

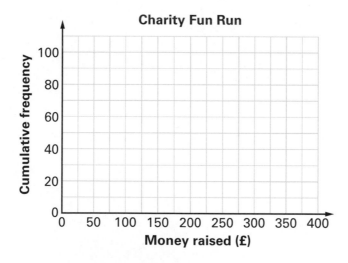

(b) Estimate the median amount of money raised.

£ _____

(c) Give the interquartile range.

_____ – _____

(d) An extra runner has raised £350.

If this is included with the others, describe the effect that this will have on the mean.

_____

_____

21. In 2011 the populations of countries in the United Kingdom were:

- England   53 012 456

- Scotland   5 295 000

- Wales   3 063 456

- N. Ireland   1 810 863

(a) What percentage of the population of the United Kingdom lived in England?

_____

(b) In 2001, the population of England was 49 138 831.

By how many percent had the population of England increased?

_____

Test Paper 1

Calculator **not** allowed

First name _____

Last name _____

Date _____

Instructions:

- The test is 1 hour long.
- Find a quiet place where you can sit down and complete the test paper undisturbed.
- You **may not** use a calculator for any question in this test.
- You will need: a pen, pencil, rubber and a ruler. You may find tracing paper useful.
- This test starts with easier questions.
- Write your answers where you see this symbol:
- Try to answer all the questions.
- The number of marks available for each question is given in the margin.
- Write all your answers **and working** on the test paper. Marks may be awarded for working.
- Check your work carefully.
- Check how you have done using pages 103–112 of the Answers and Mark Scheme.

You might need to use these formulae:

Trapezium	**Prism**
Area = $\frac{1}{2}(a + b)h$	Volume = area of cross-section × length

MAXIMUM MARK	60	ACTUAL MARK	

1. **(a)** Ewan has £60 and spends £15

What percentage of his original amount has he spent?

1 mark

_____

(b) Milly has £60 left after spending £15

What percentage of her original amount does she have left?

1 mark

_____

(c) Scott has spent 15% of his £60

How much money does he have left?

1 mark

_____

2. Here are two cuboids.

2 marks

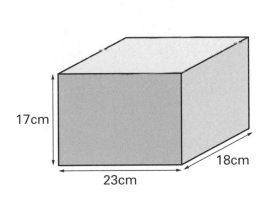

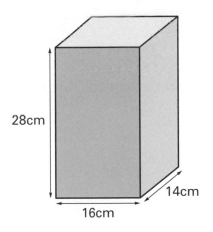

17cm

23cm

18cm

28cm

16cm

14cm

What is the difference between the volumes of the two cuboids?

Give your answer in cubic centimetres (cm^3).

_____ cm^3

SUBTOTAL

3. Here are two spinners.

Spinner A

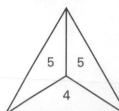

Spinner B

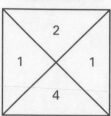

(a) Explain why it is more likely to spin a 4 on Spinner A than on Spinner B.

_____

_____

_____

Kirsty uses both spinners and totals the number from each spinner.

(b) What is the probability that Kirsty will score 10?

(c) What is the probability that Kirsty will score 6?

4. Tower A and Tower B are made with identical blocks.

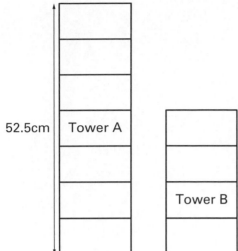

52.5cm

Tower A

Tower B

Tower A is 52.5cm tall.

(a) How tall is Tower B?

2 marks

The ratio of the length to width of each block is 2 : 1.

(b) How long is each block?

1 mark

SUBTOTAL

5. Solve:

1 mark

(a) $5x - 7 = 2x + 8$

_____

1 mark

(b) $6(y + 3) = 10(y - 3)$

_____

1 mark

6. **(a)** Ravi divides a number by 1.25 and multiplies it by 2.5.

Which number could Ravi use to multiply by that would have the same effect?

_____

Max has a recipe that makes enough for 12 people.

1 mark

(b) He multiplies the quantities of the ingredients by 0.25.

How many people will there be enough food for?

_____

1 mark

(c) He divides the quantities of the ingredients by $1\frac{1}{3}$.

How many people will there be enough food for?

_____

7. An experiment records the height of some seedlings and the number of hours of light that the seedlings receive each day.

This scatter graph shows the results after one week.

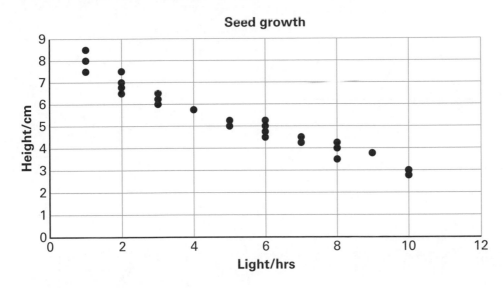

(a) Describe the correlation between the height of the seedlings and the amount of light they have received.

_____

_____

1 mark

(b) Describe the relationship between the height of the seedlings and the amount of light they receive.

_____

_____

2 marks

(c) On the graph, draw a line of best fit.

1 mark

SUBTOTAL

8. Triangle ABC is an isosceles triangle.

Lines DE and FG are parallel.

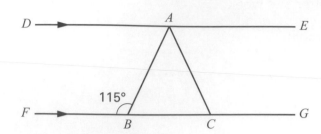

Calculate these angles.

(a) $\angle ACB =$ _____ °

(b) $\angle CAE =$ _____ °

(c) $\angle BAC =$ _____ °

9. Change these metric measures to the units shown.

(a) $6m^2 =$ _____ cm^2

(b) $1cm^3 =$ _____ mm^3

(c) A fish tank is 80cm long, 50cm wide and 60cm high.

How many litres of water can it hold?

_____ l

1 mark

1 mark

1 mark

1 mark

1 mark

1 mark

10. A gym club opened six years ago.

The club wants to know for how many months its members have been in the club.

Membership is recorded in a cumulative frequency table for the first six years.

Length of membership in years, Y	Members	Cumulative frequency
$0 < Y \leqslant 1$	40	
$1 < Y \leqslant 2$	60	
$2 < Y \leqslant 3$	110	
$3 < Y \leqslant 4$	100	
$4 < Y \leqslant 5$	80	
$5 < Y \leqslant 6$	90	

(a) Complete the cumulative frequency column.

1 mark

(b) Draw the cumulative frequency graph.

(There is a blank graph on the next page.)

2 mark

(c) Use your graph to estimate the median length of membership.

1 mark

_____

(d) Use your graph to estimate the interquartile range.

1 mark

_____

SUBTOTAL

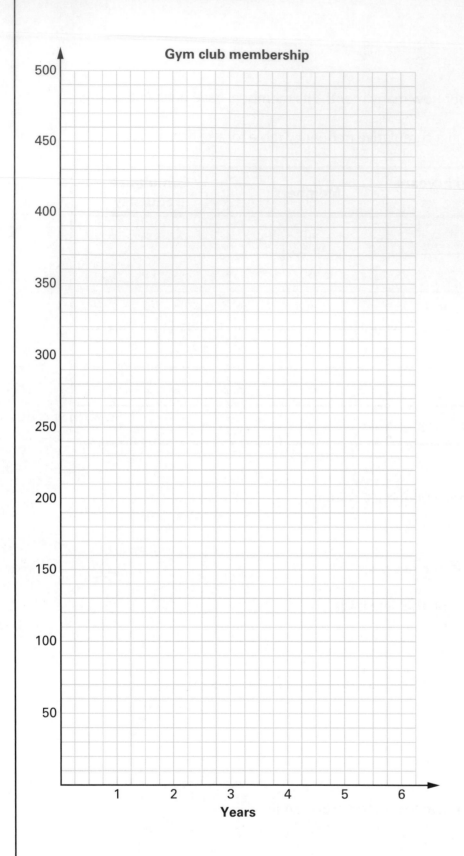

11. Here is a sequence of patterns.

Pattern number 1   Pattern number 2   Pattern number 3   Pattern number 4

(a) How many squares will be needed to make the next pattern in the sequence?

_____

1 mark

(b) Write a formula to show the number of squares in any pattern.

Begin your formula with s, where s = the number of squares and use p = the pattern number.

_____

1 mark

(c) Use your formula to find out how many squares would be needed in pattern number 48.

_____

1 mark

12. Yousef uses this spinner.

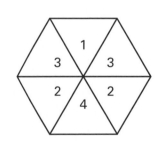

(a) Complete the table showing the theoretical probability of spinning each number.

1 mark

Number	1	2	3	4
Theoretical probability	_____	_____	_____	_____

SUBTOTAL

1 mark

(b) Complete this table showing the number of times you would theoretically spin each number.

	After 100 spins			
Number	1	2	3	4
Theoretical probability	_____	_____	_____	_____

1 mark

(c) Yousef spun his spinner 150 times. Give the number of times the spinner should theoretically land on an even number.

_____

2 marks

13. Describe fully the rotation of the shape from Shape A to Shape B.

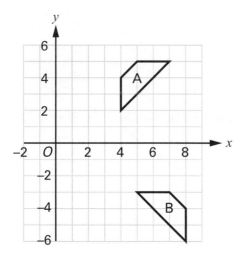

_____

_____

14. Linda sets out on a journey by car.

She leaves at 13:45 and arrives at 16:25. The journey is 120 miles.

(a) At what speed did Linda travel?

A formula for making an approximate change of miles to kilometres is:

$$Kilometres = \frac{8 \times miles}{5}$$

(b) What was Linda's speed in kilometres per hour?

(c) Linda's return journey was quicker and she travelled at 60mph.
If she left at 12:45 the next day, at what time did she arrive home?

15. The triangle ABC is enlarged by a scale factor of $\frac{1}{5}$.

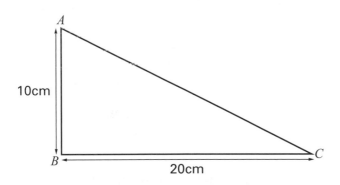

(a) What would the lengths of the sides AB and BC be?

$AB =$ _____ cm and $BC =$ _____ cm

(b) By what scale factor would the lengths of the new triangle have to be multiplied to return it to being the first triangle?

16. Two identical bags both contain three white balls and two black balls.

Maria pulls a ball from each bag.

(a) Calculate the probability that both balls will be white.

(b) Calculate the probability that both balls will be black.

Maria just uses one of the bags.

She pulls one ball from the bag, keeps the ball and then pulls another ball.

(c) Calculate the probability that both balls will be white.

(d) Calculate the probability that both balls will be black.

17. A school collects tokens from a local supermarket.

48 000 tokens are collected and need to be put into bundles of 100.

It takes 6 students $2\frac{1}{2}$ hours to count the tokens.

(a) How long would it take 10 students to count the tokens?

_____

2 marks

(b) How many students would be needed to complete the task in 60 minutes?

_____

2 marks

18. The line shows the equation $y = x - 2$.

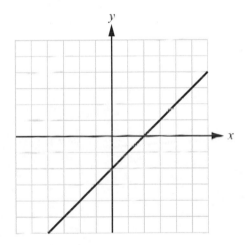

(a) Use the graph to give the value of y if $x = -3$.

 $y =$ _____

1 mark

SUBTOTAL

(b) Use the same graph to give the value of x if $y = -4$.

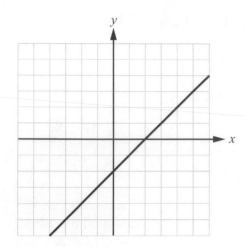

$x =$ _____

(c) Write the equation shown by this graph.

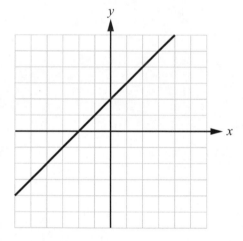

$y =$ _____

19. This cumulative frequency graph shows the ages of 160 passengers on an aeroplane.

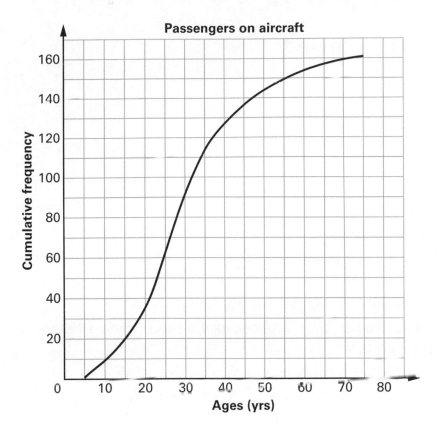

(a) Draw a box plot of the information in the cumulative graph.

2 marks

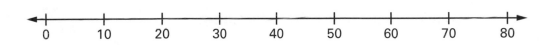

(b) Give the median age of the passengers.

1 mark

(c) Give the interquartile range of the ages of the passengers.

1 mark

SUBTOTAL

Test Paper 2

Calculator allowed

First name _____

Last name _____

Date _____

Instructions:

- The test is 1 hour long.
- Find a quiet place where you can sit down and complete the test paper undisturbed.
- You **may** use a calculator for any question in this test.
- You will need a pen, pencil, rubber, ruler, a pair of compasses and a scientific or graphic calculator.
- This test starts with easier questions.
- Write your answers where you see this symbol:
- Try to answer all the questions.
- The number of marks available for each question is given in the margin.
- Write all your answers **and working** on the test paper. Marks may be awarded for working.
- Check your work carefully.
- Check how you have done using pages 103–112 of the Answers and Mark Scheme.

You might need to use these formulae:

Trapezium	Prism
Area $= \dfrac{1}{2}(a + b)h$	Volume = area of cross-section × length

Trapezium diagram: labelled with b (top), height (h), a (bottom).

Prism diagram: labelled with length, area of cross-section.

MAXIMUM MARK	60		ACTUAL MARK	

1. **(a)** Ula earns £7.60 an hour.

She works 35 hours a week.

She is given a rise of 5%.

What will her new weekly wage be?

£ _____

2 marks

(b) Ben earns £8 an hour.

He works 30 hours a week.

He gets a pay rise of £12 a week.

What is this pay rise as a percentage?

£ _____

2 marks

2. The time in Italy is 1 hour ahead of time in the United Kingdom.

A plane leaves Manchester at 16:48.

The flight takes 2 hours and 17 minutes.

At what time does the plane land in Italy?

_____

2 marks

SUBTOTAL

3. This is the end of a building.

The roof is 6.5m high.

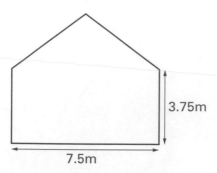

3.75m

7.5m

Calculate the area of the end wall.

Give your answer correct to 2 decimal places.

 _____ m²

4. Here are three parcels. They are the same size but weigh different amounts.

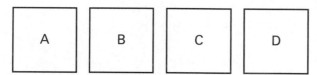

A   B   C   D

- Parcel A weighs one quarter of parcel C.

- Parcel A weighs three times parcel B.

- Parcel B weighs half of parcel D.

- Parcel D weighs 4.5kg.

What is the weight of all four parcels?

 _____ kg

5. This bar chart shows the number of times players played for their rugby club during a season.

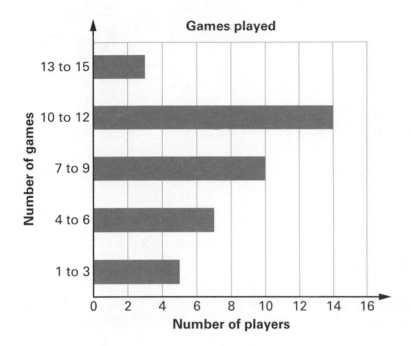

(a) How many players played in 7 or more games?

1 mark

(b) How many played at least one game?

1 mark

(c) 'At least 1 person played in all 15 games.'

Is this statement correct?

Explain your answer.

1 mark

SUBTOTAL

2 marks

6. Solve:

$$12(2t - 3) = 4t - 31$$

$t =$ _____

7. A sandwich shop makes sandwiches in batches of 20.

The owner works out the cost of each sandwich by using the formula:

$$S = \frac{F + B}{20} + 2.8$$

where S is the cost of one sandwich in pounds, F is the cost of the fillings in £ and B is the cost of the bread in £.

1 mark

(a) Find the cost of 1 cheese and tomato sandwich where the cost of the fillings is £9.25 and the cost of the bread is £4.75

£ _____

1 mark

(b) Find the cost of 12 egg and salad sandwiches, where the cost of the fillings is £5.25 and the cost of the bread is £4.75

£ _____

1 mark

(c) A sandwich sells for £3.90 If the bread cost £4.80, what was the cost of the fillings?

£ _____

8. **(a)** Calculate the area of this trapezium.

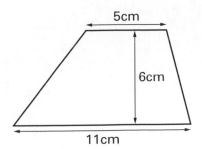

_____

1 mark

(b) This right-angled triangle has an area of 56cm^2.

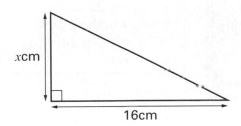

What is the height of the triangle?

_____ cm

1 mark

SUBTOTAL

9. **(a)** Write the ratio 72 : 48 : 36 in its simplest form.

_____ : _____ : _____

(b) These cans of beans come in three sizes.

One can
• Weight: 200g
• Cost: 35p

One pack of three cans
• Total weight: 750g
• Cost: £1

One can
• Weight: 450g
• Cost: 55p

Compare the unit amounts to decide which is the best value for money.

Give the unit cost per gram and tick the best-value option.

	Unit cost per gram	Best value Tick (✓)
An individual can weighing 200g		
A pack of 3 cans each weighing 250g		
An individual can weighing 450g		

10. A castle is open to visitors 5 days a week.

This pie chart shows the number of visitors.

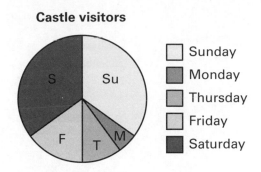

Castle visitors

There were 1000 visitors during the week.

(a) The sector showing Thursday has an angle of 36° at the centre.

How many visitors were there on Thursday?

_____

1 mark

(b) A bar chart is drawn of the same information, but two blocks are missing.

1 mark

Draw the missing blocks.

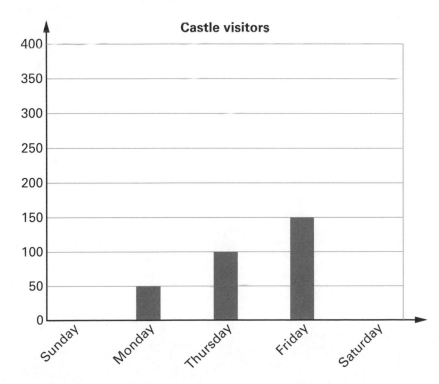

11. In a box of 200 coloured straws, there are blue, red, orange, green and yellow straws.

Straws are chosen at random.

The probabilities of taking a colour are:

- blue: $\frac{1}{8}$

- red: $\frac{1}{10}$

- orange: $\frac{1}{5}$

(a) Explain why it is impossible for there to be an equal number of green and yellow straws.

_____

_____

There are 35 green straws in the box.

Give the probability that a straw taken at random is:

(b) green

_____

(c) yellow

_____

12. This model of a steam engine is built to a scale of 1 : 80.

15.5cm

(a) How long is the real-life steam engine?

Give your answer in metres.

 _____ m

1 mark

(b) The real-life carriages are 18m long.

The model carriages are made to the same scale.

How long is a model carriage?

Give your answer in centimetres.

 _____ cm

1 mark

SUBTOTAL

13. This is a line graph showing the journey of two cars.

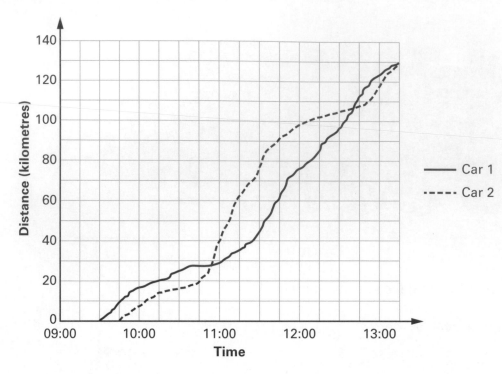

(a) After nearly 30 kilometres, Car 2 passes Car 1. At what time does this happen?

(b) What was the mean speed for Car 1's journey?

(c) Calculate the mean speed of Car 2 between 10:45 and 11:15.

(d) At about 12:40 Car 1 passes Car 2. How far had the cars travelled when this happened?

14. This design is made up from four semi-circles and a 5cm square.

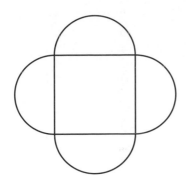

Calculate:

(a) the perimeter of the design.

Give your answer correct to 2 d.p.

(b) the area of the design.

Give your answer correct to 2 d.p.

15. Solve the following inequalities and show their solutions on the number line.

Use the correct symbols.

(a) $r + 6 < 5$

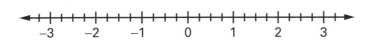

(b) $2s + 6 \geqslant 5$

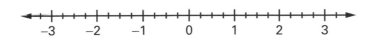

(c) $4(t - 2) \leqslant 2$

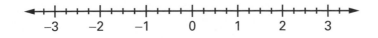

SUBTOTAL

16. A spinner is made from a regular pentagon and is numbered 1 to 5.

(a) What is the theoretical probability that a 5 is rolled?

_____ ,

Another spinner is used but is numbered differently.

The number of times that the spinner lands on 5 are recorded in this table.

(b) Complete the table to show the relative frequencies of landing on a 5.

Number of spins	10	20	30	40	50
Number of times landing on a 5	5	7	12	12	21
Relative frequency	_____	_____	_____	_____	_____

(c) How many of the sections on the second spinner would you expect to be numbered 5?

_____

17. A quadrilateral is drawn on this grid.

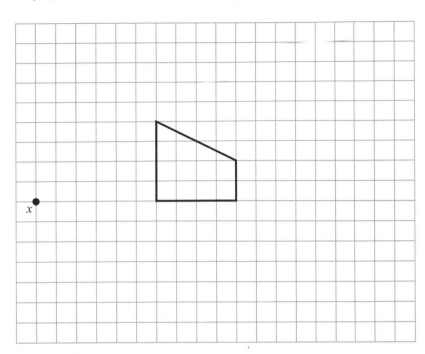

(a) Enlarge the quadrilateral by a scale factor of 1.5 about the centre of enlargement x.

Draw the quadrilateral on the grid.

2 marks

(b) Enlarge the quadrilateral by a scale factor of 0.5 about the centre of enlargement x.

Draw the quadrilateral on the grid.

2 marks

SUBTOTAL

18. In each circle, the centre of the circle is c.

Calculate the missing angles.

(a)

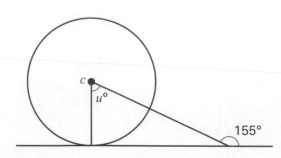

$u =$ _____ °

(b)

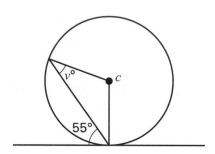

$v =$ _____ °

(c)

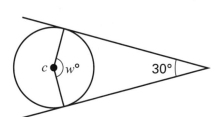

$w =$ _____ °

19. There are three bags with black and white balls.

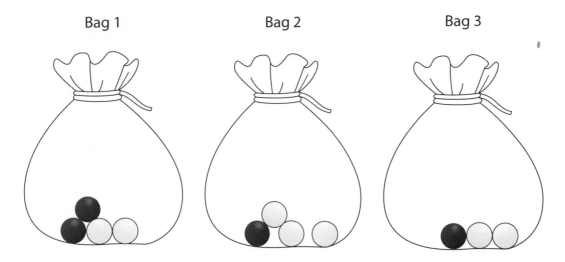

Bag 1   Bag 2   Bag 3

(a) Complete this tree diagram.

2 marks

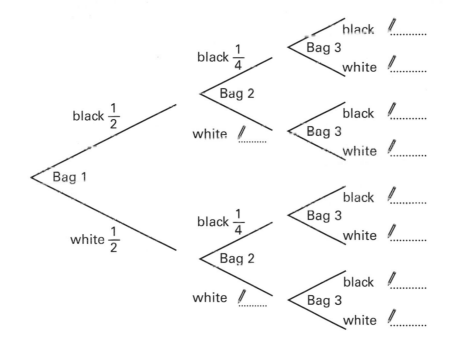

(b) What is the probability of pulling a black ball from each bag at random?

1 mark

(c) What is the probability of pulling a white ball from each bag at random?

1 mark

SUBTOTAL

1 mark

20. **(a)** Bill bought a car for £8000.

He ran the car for three years and each year the car lost 12% of its value.

What was the value of the car after the third year?

Answer to the nearest £10.

1 mark

(b) Calculate:

$$5.2(-7.2 - 8.45) \times \frac{(5.35 - 6.8)}{2^2}$$

Answer to 1 d.p.

1 mark

(c) Calculate:

$$81^{\frac{1}{2}}$$

1 mark

(d) Calculate:

$$\left(\frac{3}{4} + \frac{2}{5}\right) \div \left(\frac{3}{8} - \frac{1}{6}\right)$$

Answers and Mark Scheme

Set A, Test Paper 1

1. Signs in order should be:
$< \; < \; > \; =$ *(2 marks. 1 mark: three symbols correct)*

2. Circles joined: *(2 marks. 1 mark: 1 or 2 lines correctly drawn)*

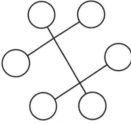

3. **(a)** 52cm   *(2 marks. 1 mark: Correct method*
e.g. 2 (15 + 11 = error or 15 + 11 + 9 + 6 + 6 + 5 = error)
   (b) 120cm^2 *(2 marks. 1 mark: Correct method*
e.g. 2 (15 + 11 = error or 15 + 11 + 9 + 6 + 6 + 5 = error)

4. **(a)** £9.60 (Do not accept £9.6)   *(1 mark)*
   (b) £21.60   *(1 mark)*

5. **(a)** 16   *(1 mark)*
   (b) 15.5   *(1 mark)*
   (c) 14   *(1 mark)*
   (d) 7   *(1 mark)*

6. **(a)** $g = 1.5$ *(1 mark)*
   (b) $10n + 2p^2 + p$ *(1 mark)*
   (c) 352   *(1 mark)*

7. 7500cm^3 *(2 marks. 1 mark: Correct method*
e.g. $\pi \times 10^2 \times 25$ = error)

8. **(a)** $2 \times 2 \times 2 \times 3$ or $2^3 \times 3$ *(1 mark)*
   (b) 12   *(1 mark)*
   (c) 120   *(1 mark)*

9. **(a)** 10   *(1 mark)*
Helpful hint Remember the number of triangles in any polygon is always two less than the number of sides.
   (b) 1800°   *(1 mark)*
   (c) 30°   *(1 mark)*
Helpful hint The exterior angle of any regular polygon is 360° ÷ by the number of sides of the polygon.

10. **(a)** Explanation could use the fact that: *(1 mark)*
   • the probability of there being a black counter is 0.1 and 25 × 0.1 does not equal a whole number.

• the probability of a blue counter is 0.4 and there are 12 blue counters, therefore there are 30 counters.
   (b) $1 - (0.2 + 0.4 + 0.1) = 0.3$ *(1 mark)*
   (c) 9 white counters   *(1 mark)*

11. **(a)** 9cm   *(1 mark)*
   (b) 72cm, 96cm, 108cm   *(2 marks. 1 mark: Two correct lengths)*
   Answers can be given in any order.

12. **(a)** 1: 5 expected successes
   2: 2 expected successes
   3: 9 expected successes   *(2 marks. 1 mark: two correct answers)*
   (b) No. Explanation needs to refer to the table: the expected probabilities are not the same, which they would be on a fair dice.   *(1 mark)*

13. **(a)** Total of frequencies = 20   *(1 mark)*
   Score × frequency column completed from top to bottom as follows: 20, 38, 54, 119, 64, 45, 340   *(2 marks. 1 mark: 4 or more correct multiplications)*
   (b) 58   *(1 mark)*

14. **(a)** The pattern of the lines should be: *(2 marks. 1 mark: Two or three lines correctly drawn)*

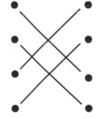

   (b) $y = 4x - 5$ *(1 mark)*

15. **(a)** $\frac{3}{5}$ *(1 mark)*
   (b) 225ml   *(1 mark)*
   (c) 250ml   *(2 marks. 1 mark: Correct method e.g. Sally has 200ml of pear juice and 300ml of apple juice. To find Yasmin's juice, 300 ÷ 2 = 150, 150 × 3 = error, error – 200)*

16. **(a)** **(iii)** $a(b - d) + bc$ *(1 mark)*
   (b) $b = \frac{P - 2a}{2}$ *(1 mark)*

17. **(a)** 12cm   *(1 mark)*
   (b) 60cm^2 *(1 mark)*

(c) 52° *(1 mark)*

Helpful hint *The two chords AE and BE will always form a right angle when they meet on the circumference and join the ends of the diameter.*

18. (a) 1 : 8 *(2 marks. 1 mark: Correct method e.g. $6 \times 4 \times 2$ = error 1, $12 \times 8 \times 4$ = error 2, error 2 ÷ error 1)*

(b) 1 : 64 *(2 marks. 1 mark: Correct method e.g. $6 \times 4 \times 2$ =error 1, $24 \times 16 \times 8$ = error 2, error 2 ÷ error 1)*

19. (a) $0.1 \times 0.1 = 0.01$ *(1 mark)*
(b) $1 - 0.01 = 0.99$ *(1 mark)*
(c) $0.5 \times 0.4 = 0.2$ *(1 mark)*

Helpful hint *Remember to multiply probabilities to find the outcome of two independent events.*

20. (a) 1.5×10^9 *(1 mark)*
(b) 1.5×10^4 *(1 mark)*

Helpful hint *Re-arrange expressions, e.g. $(3 \times 10^5) \times (5 \times 10^3) = 3 \times 5 \times 10^5 \times 10^3$ Indices are added when multiplying*
$$= 15 \times 10^8$$
Indices are subtracted when dividing.

21. 2.5cm *(1 mark)*

Set A, Test Paper 2

1. (a) −2, −9, −16 *(1 mark)*
(b) 12, 20, 36 *(1 mark)*
(c) 5, 8, 11 *(1 mark)*

2. (a) 5 : 3 : 4 *(1 mark)*
(b) $\frac{1}{4}$ *(1 mark)*

3. (a) $\frac{3}{5}$ or 0.6 *(1 mark)*
(b) $\frac{3}{8}$ or 0.375 *(1 mark)*
(c) $\frac{1}{5}$ or 0.2 *(1 mark)*

4. (a) 19:02 *(1 mark)*
(b) 15:58 *(1 mark)*
(c) 19:32 *(1 mark)*

Helpful hint *On a timetable, empty spaces mean the train doesn't stop.*

5. (a) 4 edges *(1 mark)*
(b) 352cm^2 *(2 marks. 1 mark: Correct method e.g. $2 (12 \times 8) = 2 (12 \times 4) + 2 (8 \times 4)$)*

6. (a) 102.10cm^2 *(2 marks. 1 mark: Correct method e.g. $\frac{(\pi \times 9^2)}{2} - \frac{(\pi \times 4^2)}{2}$)*
(b) 50.84cm *(2 marks. 1 mark: Correct method e.g. $\frac{(\pi \times 18)}{2} + \frac{(\pi \times 8)}{2} + (2 \times 5)$)*

7. (a) $4s^2 - 20s$ *(1 mark)*
(b) $r^2 + 13r + 42$ *(1 mark)*
(c) $5(2t + 5)$ *(1 mark)*

8. (a) 600 *(1 mark)*
(b) £32.50 *(1 mark)*

Helpful hint *When you read from a calculator and the answer refers to money, two numbers are always needed after the decimal point. So, £32.5 would be a wrong answer.*

9. (a) 0.2 or $\frac{1}{5}$ or $\frac{12}{60}$ *(1 mark)*

(b) $0.\dot{3}$ or $\frac{1}{3}$ or $\frac{20}{60}$ *(1 mark)*

(c) No. Explanation needs to show that it is possible to pick a jumper that is red and large at the same time. *(1 mark)*

Helpful hint *Mutually exclusive events are ones that can't happen at the same time: e.g. tossing a coin and the result being both a head and a tail.*

10. (a) £26.65 *(1 mark)*
(b) The following should be circled:
• £11.90 (1st class, 1kg–2kg)
• £8.90 (2nd class, 2kg–5kg) *(2 marks. 1 mark: Correct method e.g. $26.40 - 5.60$)*

11. (a) 7 *(2 marks. 1 mark: Works to correct equation e.g. $5x + 13 = 45$)*
(b) 5 *(2 marks. 1 mark: Works to correct equation e.g. $16x - 8 = 9x + 27$)*

12. 78.78 *(1 mark)*

13. Second box indicated only. *(1 mark)*

14. A + H and E + D *(1 mark)*

15. (a) 20% *(1 mark)*
(b) 6% *(2 marks. 1 mark: Correct method e.g. $(13.25 \times 12) - 150$ = error; error ÷ 150 × 100)*
(c) 40% *(2 marks. 1 mark: Correct method e.g. 40×3 = error 1, error 1 − 72 = error 2, error 2 ÷ error 1 × 100)*

Helpful hint *When finding percentages, remember you need to make a fraction. In this case, the whole price is £120. The price has been reduced by £48 (120 – 72)*
So the fraction is $\frac{48}{120}$ and the calculation becomes $48 \div 120 \times 100 = 40$

16. The order of these answers is not important.
(a) Any two numbers that total 23, e.g. 1 and 22, 2 and 21, 3 and 20 … *(1 mark)*

Helpful hint *If the mean is 16 and there are 7 numbers, the total of the numbers must be 112 (16 × 7). The given numbers total 89, so the other two numbers must total 23 (112 – 89).*

    (b) 2 or 24   *(1 mark)*

Helpful hint *The range is 16, which could mean either*
- *6 (the lowest number) + 18 = 24*
- *20 (the highest number) – 18 = 2*

    (c) 9 and 11 or 9 and 12 (ignoring the order)   *(1 mark)*

Helpful hint *Arrange the numbers in order –*
         9   9   10   10   12
Another 9 must be added to make the mode 9. The range is 3, but the other number cannot be 9 since this would make the median 9, and it cannot be 10 as this would make the mode 9.5. So the other number could be 11 or 12.

17. (a) 4   *(2 marks. 1 mark: Forms correct equation e.g. 4a + 3 = 5a – 1)*

    (b) $b = 5$ and $c = 4$ *(2 marks. 1 mark: Works to correct equation e.g. 13b = 65)*

    (c) **(i)** $P = 10x - 17$ *(1 mark)*
        (ii) 23 cm, 17 cm, 13 cm   *(1 mark)*

18. (a) $0.1\dot{6}$ or $\frac{1}{6}$ or $\frac{15}{90}$ *(1 mark)*

    (b) $0.\dot{3}$ or $\frac{1}{3}$ or $\frac{30}{90}$ *(1 mark)*

    (c) The probability of it not raining on the first day is 0.8   *(2 marks. 1 mark: Correct method e.g. 0.8 × 0.8)*
    The probabilities of 0.2 and 0.8 also apply on the second day and should be written in spaces provided.
    P(no rain for 2 days) = 0.8 × 0.8 = 0.64

19. (a) Accept an answer ⩾3.6 km and ⩽3.9 km   *(1 mark)*

    (b) Accept answers ⩾ 2.6, ⩽ 2.9 and ⩾5.6, ⩽ 5.9   *(2 marks. 1 mark for each outlier)*
    Accept also answers in the range of 2.7–3.3 inclusive.

20. (a) $18d^{-5}$ *(1 mark)*
    (b) $3c^9$ *(1 mark)*
    (c) $5a^2b^2$ *(1 mark)*

Set B, Test Paper 1

1. (a) 0.8m or 80cm   *(1 mark)*
    (b) 53 DVDs   *(2 marks. Accept follow through from 1(a) 1 mark: Correct method e.g. 80 ÷ 1.5 = error)*

Helpful hint *Remainders here do not apply. You can't put one-third of a DVD on a shelf.*

2. 40.18 (to 2 d.p.)   *(2 marks. 1 mark: Correct method with no more than one arithmetical error)*

3. (a) 18°   *(1 mark)*
    (b) Explanation should indicate that isosceles triangles have two equal angles.
    Do **not** accept: They don't look like it.   *(1 mark)*
    (c) • 28°, 76°, 76°
        • 52°, 52°, 76°   *(2 marks. 1 mark for each set of correct angles)*

4. (a) $(2a - 3)(a + 3) = 2a^2 + 3a - 9$ *(1 mark)*
    $(2a - 3)(a + 3)$ is not sufficient for the mark.
    (b) 340cm or 3.4m   *(1 mark)*

5. (a) 180 ± 2   *(1 mark)*
Helpful hint *The angle of this sector is a right angle.*
    (b) Cycle   *(1 mark)*
Helpful hint *12.5% is $\frac{1}{8}$*

6. (a) 36   *(1 mark)*
    (b) $\frac{1}{12}$ or $\frac{5}{60}$ *(1 mark)*
    (c) 72°   *(1 mark)*
Helpful hint *12 out of 60 = $\frac{1}{5}$*
$\frac{1}{5}$ of 360° = 72°

7. (a) 8cm   *(2 marks. 1 mark: Correct method e.g. 4 ÷ 50 000 × 1000 × 100 = error)*
    (b) 12.5km   *(2 marks. 1 mark: Correct method e.g. 25 × 50 000 ÷ 100 ÷ 1000 = error)*

8. 1133.6   *(1 mark)*

9. (a) ☑ ☐ ☐
    (b) ☐ ☐ ☑
    (c) ☐ ☑ ☐
    (d) ☐ ☐ ☑   *(2 marks. 1 mark for three correct answers)*

10. (a) $M = x + 3x + 4x = 8x$ *(1 mark)*
    (b) Mark has 35 marbles and Barry has 105 marbles.   *(1 mark)*

11. (a) 85%   *(1 mark)*
    (b) 14   *(1 mark)*
    (c) Explanation should show that $\frac{18}{25} = 72\%$ and that 72% < 85%   *(1 mark)*

12. (a) 2kg   *(1 mark)*
    (b) 0.8kg   *(2 marks. 1 mark: Correct method e.g. Nitrogen is 25% (1/4) of the total amount of fertiliser. Therefore 200g × 4 = 800g)*

13. (a) 560   *(1 mark)*
    (b) 300   *(1 mark)*

14. (i) Is not congruent.
    (ii) Is not congruent.

Helpful hint *The three angles are the same, but this has no effect on the length of the sides.*

    (iii) Are congruent.
        Condition of congruency:
        side, angle, side (SAS) *(3 marks. 1 mark for correct ticks; 1 mark for correct condition for congruency)*

15. (a) 80   *(2 marks. 1 mark: Correct method e.g. 1 ÷ 0.125 = error, error × 10 = error)*

    (b) The teacher did not select: *(2 marks. 1 mark: 3 correct answers)*
      • Large red pentagons
      • Medium yellow pentagons
      • Medium blue pentagons
      • Large blue pentagons

16. (a) $x = 4$ *(1 mark)*
    (b) $x = 4, y = 3$ *(1 mark)*
    (c) $f = \dfrac{d^2}{e-3}$ Accept $f = \dfrac{d^2}{e+4-7}$ *(1 mark)*

17. (a) 21   *(1 mark)*

(b) 39   *(1 mark)*

18. (a) GF = 5cm   *(1 mark)*
    (b) CF = 12cm   *(1 mark)*
    (c) BC = 3cm   *(1 mark)*

Helpful hint *Add the lengths you know to the diagram. Use ratio to find the other lengths.*

19. (a) 67cm^2 *(2 marks; 1 mark: Finding correct triangle 3.5cm × 2cm; 1 mark: Correct method e.g.* $92 - 4 \times \dfrac{(3.5 \times 2)}{2}$*)*

    (b) 1005cm^3 *(2 marks. 1 mark: Answer from 19 (a) × 15 = error)*

20. (a) 0.44 or $\dfrac{11}{25}$ or $\dfrac{55}{125}$ *(1 mark)*

    (b) 0.36 or $\dfrac{9}{25}$ or $\dfrac{45}{125}$ *(1 mark)*

    (c) 5 students   *(1 mark)*

    (d) Explanation should show that 0.1 of 125 would be 12.5 or $12\frac{1}{2}$ and it would be impossible to have 0.5 or $\frac{1}{2}$ of a student.   *(1 mark)*

21. (a) The graph shows a positive correlation.   *(1 mark)*

    (b)

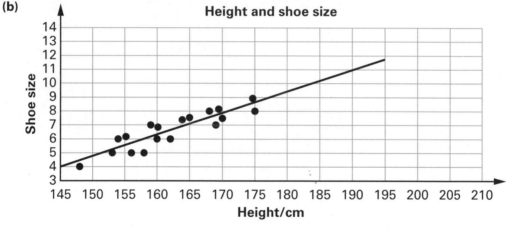

(1 mark)

    (c) 11 or 12 (in the context of the question, these are the nearest possible sizes)   *(1 mark)*
    (d) No. Explanation to include that these are only statistical trends.   *(1 mark)*

Set B, Test Paper 2

1. (a) 20°   *(1 mark)*
    (b) 130°   *(1 mark)*
    (c) 155°   *(2 marks. 1 mark: Correct method e.g. (180 − 115) + 90 = error 1; 180 − error 1 = error 2; 180 − error 2)*

2. (a) 25 downloads   *(1 mark)*
    (b) 14 : 5   *(1 mark)*

Helpful hint *Ben has 25 downloads and 70 CDs (35 × 2). The ratio is 25 : 70; simplified, this is 5 : 14.*

3. (a) 0.6 or $\dfrac{3}{5}$ *(1 mark)*

    (b) 0.5 or $\dfrac{1}{2}$ or $\dfrac{2}{4}$ *(1 mark)*

4. (a) 10   *(1 mark)*
    (b) 1 or 9   *(2 marks. 1 mark for each answer)*

5. (a) 14   *(1 mark)*
    (b) $b = 2.25$ (Accept $2\frac{1}{4}$ or $\frac{9}{4}$)   *(1 mark)*

6. (a) £22.40   *(1 mark)*
    £22.4 is **not** an acceptable answer.
    (b) £35   *(1 mark)*

7. You should use a pair of compasses and show how you draw your answers. Add a pair of construction marks on the horizontal line, both the same distance from the centre point.

(a)

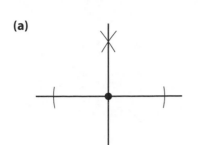

(1 mark)

(b)

(1 mark)

8. **(a)** 2 : 3   *(2 marks. 1 mark: 30 : (75 – 30 = error))*
    (b) 4 : 9   *(2 marks. 1 mark: 75 – 30 = error 1,*
30 × 23 = error 2, error 2 : error 1)

9. **(a)** $12\frac{13}{24}$ *(1 mark)*

    (b) $2\frac{11}{15}$ *(1 mark)*

10. **(a)** (0, 2) (3, 2) (–3, –5) (3, –5)   *(1 mark)*
    (b) (0, 2) (–3, 2) (–3, –5) (3, –5)   *(1 mark)*

11. $\frac{1}{7}$ *(2 marks. 1 mark: Correct method e.g. use*
of algebra, let 1 = x, so x + 2x + 4x = 7x)
Helpful hint *Think of spinning a 1 as x, then spinning a 2 will*
be 2x and spinning a 3 will be 4x (twice as likely as spinning a 2).
So, the total is x + 2x + 4x = 7x; x is $\frac{1}{7}$ of 7x

12.   *(2 marks. 1 mark for 9–11 correctly completed cells)*

	Male	Female
$10 < A \leqslant 20$	1	1
$20 < A \leqslant 30$	4	5
$30 < A \leqslant 40$	4	5
$40 < A \leqslant 50$	3	2
$50 < A \leqslant 60$	2	0
$60 < A \leqslant 70$	1	2

Helpful hint *Don't forget the 0.*

13. (a) 122.72cm^2 *(2 marks. 1 mark: Use of correct*
formula, πr^2)
    (b) 134.13cm^2 *(2 marks. 1 mark: Allow follow through*
from 13(a) or Correct method e.g. 625 – 4 × area of circle)

14. (a) $5(x – 5)$ *(1 mark)*
    (b) Accept $2b(2ab + c)$ or
       $2(2ab^2 + bc)$ or $b(4ab + 2c)$ *(1 mark)*
    (c) $3d$ *(1 mark)*

15. (a)   *(2 marks. 1 mark: 2 or 3 correct answers)*

	Per 120g	Per 50g
Protein	9g	3.75g
Carbohydrates of which sugars	79.4g **20.9g**	**33.1g** 8.7g
Fat of which saturates	2.1g 0.6g	**0.9g** 0.25g
Fibre	**4.3g**	1.8g

(b) (i) 67.5g
    (ii) 4.5g *(2 marks. 1 mark for each correct answer)*

16. (a) Explanation should show that the probabilities
       given total 1.   *(1 mark)*
    (b) £28.95   *(2 marks. 1 mark: Correct identification of*
number of 3 or 4 coins, £1 (18), 50p (15), 20p (12), 10p (6), 5p (9))
Helpful hint *Having used the probability to work out*
the number of coins, don't forget to turn it into money.
E.g. 15 50p coins = £7.50.

17. (a) $\frac{1}{8}$ or $\frac{4}{32}$ *(1 mark)*
    (b)   *(2 marks. 1 mark: 3 or 4 correctly named*
sectors or correct sectors but not named)

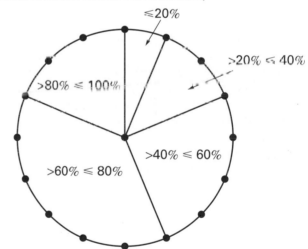

    (c) 135°   *(1 mark)*

18. (a) 6cm, 8cm, 10cm   *(1 mark)*
    (b) 2.7cm^2 *(2 marks. 1 mark: Correct method e.g.*
(1.5 × 3.6) ÷ 2 = error)
    (c) 20cm   *(1 mark)*

19. (a) 1, 16, 41   *(1 mark)*
    (b) $y = x^2 + 2$ *(1 mark)*
Helpful hint *You should recognise this type of curve as*
representing a quadratic equation. This means it involves
square numbers.
Find the coordinates you can find easily:
(0, 2) (1, 3) (2,6) (3, 11)
(0, 2) is useful; if x = 0, then x^2 also equals 0, 2 is added in
the y-axis.

(c) *(1 mark)*

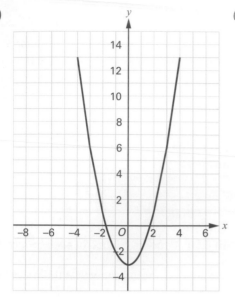

(a) *(1 mark)*

Charity Fun Run

(b) In the range of £155–£170 inclusive *(1 mark)*

(c) In the range £110 (±5) – £230 (±5) inclusive *(1 mark)*

(d) Explanation should show that as £350 is above the mean, so the mean will increase. *(1 mark)*

20.

Amount raised per person (n)	Frequency	Cumulative frequency
$0 < n \leqslant 50$	5	**5**
$50 < n \leqslant 100$	12	**17**
$100 < n \leqslant 150$	28	**45**
$150 < n \leqslant 200$	20	**65**
$200 < n \leqslant 250$	18	**83**
$250 < n \leqslant 300$	12	**95**
$300 < n \leqslant 350$	4	**99**
$350 < n \leqslant 400$	1	**100**

21. (a) 83.9% (to one d.p.) *(1 mark)*

(b) 7.9% (to one d.p.) *(1 mark)*

Set C, Test Paper 1

1. (a) 25% *(1 mark)*

(b) 80% *(1 mark)*

(c) £51 *(1 mark)*

2. 766cm³ *(2 marks. 1 mark: Correct method e.g. $(23 \times 18 \times 17) - (28 \times 16 \times 14)$ = error)*

3. (a) Explanation should show that the probability of spinning a 4 on Spinner 1 is $\frac{1}{3}$ and spinning a 4 on Spinner 2 is $\frac{1}{4}$ **and** $\frac{1}{3} > \frac{1}{4}$ Accept: The angle at centre of Spinner 1 is 120°; at the centre of Spinner 2 the angle is 90° and 120° > 90°. *(1 mark)*

(b) 0 (It is impossible to total 10 from the numbers on the spinners.) *(1 mark)*

(c) 0.42 (2 d.p.) or $\frac{5}{12}$ *(1 mark)*

4. (a) 30cm *(2 marks. 1 mark: Correct method e.g. $52.5 \div 7 \times 4$ = error)*

(b) 15cm *(1 mark)*

5. (a) $x = 5$ *(1 mark)*

(b) $y = 12$ *(1 mark)*

6. (a) 2 *(1 mark)*

(b) 3 *(1 mark)*

(c) 9 *(1 mark)*

7. (a) There is a negative correlation. *(1 mark)*

(b) The less light the seedlings receive, the higher they grow.
or
The more light the seedlings receive, the less they grow. *(2 marks. 1 mark: Incomplete statement e.g. The less light the seedlings receive affects their growth.)*

(c) *(1 mark)*

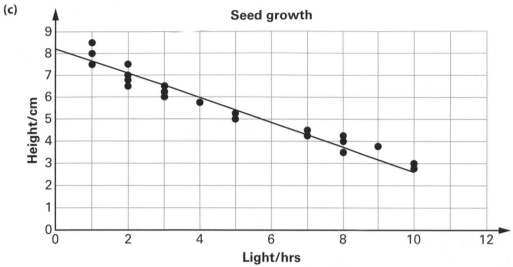

8. (a) 65° *(1 mark)*
(b) 65° *(1 mark)*
(c) 50° *(1 mark)*

9. (a) 60 000cm^2 *(1 mark)*
(b) 1000mm^3 *(1 mark)*
(c) 240l *(1 mark)*

10. (a) *(1 mark)*

Length of membership in years, Y	Members	Cumulative frequency
$0 < Y \leq 1$	40	**40**
$1 < Y \leq 2$	60	**100**
$2 < Y \leq 3$	110	**210**
$3 < Y \leq 4$	100	**310**
$4 < Y \leq 5$	80	**390**
$5 < Y \leq 6$	90	**480**

(b) *(2 marks. 1 mark for a correct graph using (incorrect) data from 10 (a))*

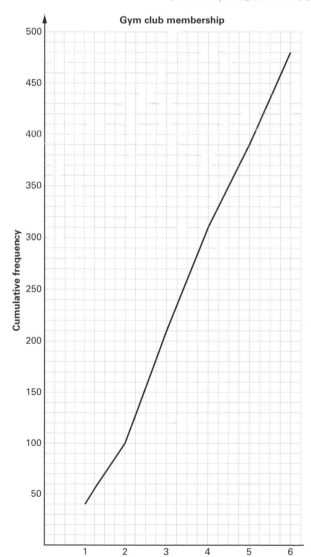

(c) In the range $\geqslant 3.1$, $\leqslant 3.4$ *(1 mark)*

(d) In the range from $\geqslant 2.1$ years, $\leqslant 2.4$ years to > 4 years, < 4.25 years *(1 mark)*

11. (a) 16 squares *(1 mark)*

(b) $s = 3p + 1$ *(1 mark)*

Helpful hint *The pattern is growing by 3 at each step, so 3 must be the multiplier in the formula.*
Looking back to pattern number 1, the first multiple of 3 is 3. There is 1 extra square, which gives + 1.

(c) 145 squares *(1 mark)*

12. (a) Probability of spinning *(1 mark)*
- 1: $\frac{1}{6}$ or $0.1\dot{6}$
- 2: $\frac{1}{3}$ or $\frac{2}{6}$ or $0.\dot{3}$
- 3: $\frac{1}{3}$ or $\frac{2}{6}$ or $0.\dot{3}$
- 4: $\frac{1}{6}$ or $0.1\dot{6}$

(b) Theoretical probability of spinning each number: *(1 mark)*
- 1: 16.7 or $16\frac{2}{3}$ or 17
- 2: 33.3 or $33\frac{1}{3}$ or 33
- 3: 33.3 or $33\frac{1}{3}$ or 33
- 4: 16.7 or $16\frac{2}{3}$ or 17

(c) 75 *(1 mark)*

13. A clockwise rotation of 90° about the point (2, −1) or an anti-clockwise rotation of 270° about the point (2, −1) *(2 marks. 1 mark: identification of point (2, −1); 1 mark: correct description of rotation)*

14. (a) 45mph *(1 mark)*

(b) 72km/h *(1 mark)*

(c) 14:45 *(1 mark)*

15. (a) $AB = 2$cm, $BC = 4$cm *(2 marks. 1 mark for each correct length)*

(b) 5 *(1 mark)*

16. (a) 0.36 or $\frac{9}{25}$ *(1 mark)*

Helpful hint *Remember to multiply the probabilities to find the probability of two events. In this case:*
- 0.6×0.6
- $\frac{3}{5} \times \frac{3}{5}$

(b) 0.16 or $\frac{4}{25}$ *(1 mark)*

(c) 0.3 or $\frac{3}{10}$ *(1 mark)*

Helpful hint *Remember: having kept one ball, the probability of the second ball being white is now $\frac{1}{2}$ or $\frac{2}{4}$.*

(d) 0.1 or $\frac{1}{10}$ *(1 mark)*

17. (a) $1\frac{1}{2}$ hours or 90 minutes *(2 marks. 1 mark: Correct method e.g. $150 \times \frac{3}{5} = $ error)*

Helpful hint *The number of tokens is a distractor and not relevant.*

The key facts are 6 students take 150 minutes ($2\frac{1}{2}$ hours).

The question asks how long 10 students will take.

The number of students has increased by $1\frac{2}{3}$ or $\frac{5}{3}$

To find the time taken we need to increase the time taken by the reciprocal of $\frac{5}{3}$.

You can find this by inverting $\frac{5}{3}$, this is $\frac{3}{5}$.

$150 \times \frac{3}{5} = 90$ minutes or $1\frac{1}{2}$ hours.

(b) 15 students *(2 marks. 1 mark: Correct method e.g. $6 \times \frac{5}{2} = $ error)*

Helpful hint *The time has been multiplied by $\frac{2}{5}$, so we need to multiply the number of students by the reciprocal of $\frac{2}{5}$; which is $\frac{5}{2}$ or 2.5; $6 \times 2.5 = 15$*

18. (a) $y = -5$ *(1 mark)*

(b) $x = -2$ *(1 mark)*

(c) $y = x + 2$ *(1 mark)*

19. (a) *(2 marks)*

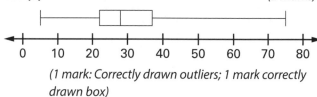

(1 mark: Correctly drawn outliers; 1 mark correctly drawn box)

(b) 28 years *(1 mark)*

(c) 15 years *(1 mark)*

Set C, Test Paper 2

1. (a) £279.30 *(2 marks. 1 mark: Correct method e.g. $7.6 \times 35 = $ error; error $+ \frac{5}{100}$ of error)*
£279.3 is **not** acceptable.

(b) 5% *(2 marks. 1 mark: Correct method e.g. $12 \div (8 \times 30) \times 100$)*

2. 20:05 *(2 marks. 1 mark: 18:05 or 19:05)*

3. 38.44m^2 *(2 marks)*
(1 mark: Correct method e.g.
$6.5 - 3.75 = $ error 1
$(7.5 \times 3.75) + \left(\frac{7.5 \times \text{error 1}}{2} \right) = $ error 2

4. 40.5kg *(2 marks)*
  (1 mark: Two correctly identified weights:)
  • A = 6.75kg
  • B = 2.25kg
  • C = 27kg
Helpful hint Start with the given fact:
D = 4.5kg.
B is half of D.
B = 4.5 ÷ 2 = 2.25kg.
A is three times B.
A = 2.25 × 3 = 6.75kg.
A is a quarter of C.
C = 6.75 × 4 = 27kg.

5. **(a)** 27 players *(1 mark)*
  (b) 39 players *(1 mark)*
  (c) This statement is incorrect, so NO. Explanation needs to show that the 13–15 group only indicates players within that group, not the specific number of games. E.g. all three players in this section may have played 13 games. *(1 mark)*

6. $t = 0.25$ or $\frac{1}{4}$ *(2 marks. 1 mark: Correct expression e.g. $24l - 36 = 4t - 31$)*

7. **(a)** £3.50 *(1 mark)*
  £3.5 is **not** an acceptable answer.
  (b) £39.60 *(1 mark)*
  £39.6 is **not** an acceptable answer.
  (c) £17.20 *(1 mark)*
  £17.2 is **not** an acceptable answer.

8. **(a)** 48cm^2 *(1 mark)*
  (b) 7cm *(1 mark)*

9. **(a)** 6 : 4 : 3 *(1 mark)*
  (b)

	Best value Tick (✓)	Unit cost per gram
An individual can weighing 200g		0.175p
A pack of 3 cans each weighing 250g		0.13˙p
An individual can weighing 450g	✓	0.12˙p

  (2 marks. 1 mark for correct best value indicated. 1 mark for three correct unit costs per gram)

10. **(a)** 100 visitors *(1 mark)*
  (b) The blocks for Sunday and Saturday should both be drawn to 350. *(1 mark)*

11. **(a)** Blue straws = $\frac{1}{8}$ of 200 = 25; red straws = $\frac{1}{10}$ of 200 = 20; orange straws = $\frac{1}{5}$ of 200 = 40
  25 + 20 + 40 = 85
  200 − 85 = 115, and 115 cannot be divided equally between green and yellow straws.
  (2 marks. 1 mark: Correct method e.g. $1 - (\frac{1}{8} + \frac{1}{10} + \frac{1}{5})$ = error. Error × 200)
  (b) $\frac{7}{40}$ or 0.175 *(1 mark)*
  (c) $\frac{2}{5}$ or 0.4 *(1 mark)*

12. **(a)** 12.4m *(1 mark)*
  (b) 22.5cm *(1 mark)*

13. **(a)** 10:50–10:55 *(1 mark)*
  (b) 34.7km/h *(1 mark)*
  (c) 114km/h ±4km/h *(1 mark)*
  (d) 107 ±2km *(1 mark)*

14. **(a)** 31.42cm *(2 marks. 1 mark: Correct method e.g. $2 \times \pi d$)*
  (b) 64.27cm^2 *(2 marks. 1 mark: Correct method e.g. $2 \times \pi r^2 + 5^2$)*

15. **(a)** *(1 mark)*
  −1
  (b) *(1 mark)*
  −0.5
  (c) *(1 mark)*
  2.5

Helpful hint Remember that the symbols for inequalities consist of two parts:
• An arrow – this tells you the direction of the values; they may be increasing (to the right) or decreasing (to the left)
• A circle – this tells you whether the number indicated is included in the inequality; a shaded circle indicates it is; an empty circle indicates that it is not.

16. **(a)** 0.2 or $\frac{1}{5}$ *(1 mark)*
  (b) *(1 mark)*

Number of spins	10	20	30	40	50
Number of times landing on a 5	5	7	12	12	21
Relative frequency	0.5	0.35	0.4	0.3	0.42

  (c) 2 *(1 mark)*

17. (a)

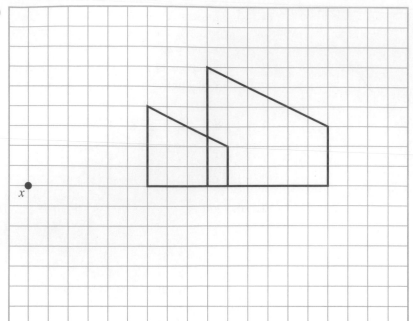

(2 marks. 1 mark: Correct enlargement incorrectly placed)

(b)

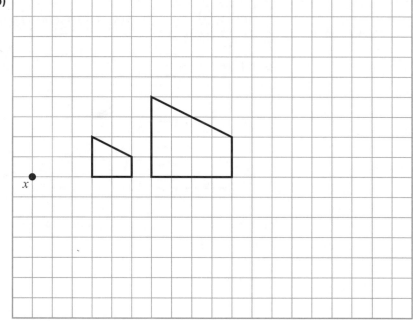

(2 marks. 1 mark: Correct enlargement incorrectly placed)

18. (a) $u = 65°$ *(1 mark)*
   (b) $v = 35°$ *(1 mark)*
   (c) $w = 150°$ *(1 mark)*

19. (a) For Bag 2 *(2 marks. 1 mark: Correct probabilities for Bag 2; 1 mark: Correct probabilities for Bag 3)*
   • white $= \frac{3}{4}$

   For Bag 3

   • black $= \frac{1}{3}$

   • white $= \frac{2}{3}$

   (b) $\frac{1}{24}$ *(1 mark)*

   (c) $\frac{1}{4}$ or $\frac{6}{24}$ *(1 mark)*

20. (a) £5450 *(1 mark)*
   (b) 29.5 *(1 mark)*
   (c) 9 *(1 mark)*
   (d) $5\frac{13}{25}$ or 5.52 *(1 mark)*